**New Directions for
Institutional Research**

Paul D. Umbach
EDITOR-IN-CHIEF

J. Fredericks Volkwein
ASSOCIATE EDITOR

Attracting and
Retaining Women
in STEM

Joy Gaston Gayles
EDITOR

Number 152 • Winter 2011
Jossey-Bass
San Francisco

ATTRACTING AND RETAINING WOMEN IN STEM
Joy Gaston Gayles (ed.)
New Directions for Institutional Research, no. 152
Paul D. Umbach, Editor-in-Chief

NEW DIRECTIONS FOR INSTITUTIONAL RESEARCH (ISSN 0271-0579, electronic ISSN 1536-075X) is part of The Jossey-Bass Higher and Adult Education Series and is published quarterly by Wiley Subscription Services, Inc., A Wiley Company, at Jossey-Bass, One Montgomery Street, Suite 1200, San Francisco, California 94104-4594 (publication number USPS 098-830). Periodicals Postage Paid at San Francisco, California, and at additional mailing offices. POSTMASTER: Send address changes to New Directions for Institutional Research, Jossey-Bass, One Montgomery Street, Suite 1200, San Francisco, California 94104-4594.

SUBSCRIPTIONS cost $109 for individuals and $297 for institutions, agencies, and libraries in the United States. See order form at end of book.

EDITORIAL CORRESPONDENCE should be sent to Paul D. Umbach, Leadership, Policy and Adult and Higher Education, North Carolina State University, Poe 300, Box 7801, Raleigh, NC 27695-7801.

New Directions for Institutional Research is indexed in *Academic Search* (EBSCO), *Academic Search Elite* (EBSCO), *Academic Search Premier* (EBSCO), *CIJE: Current Index to Journals in Education* (ERIC), *Contents Pages in Education* (T&F), *EBSCO Professional Development Collection* (EBSCO), *Educational Research Abstracts Online* (T&F), *ERIC Database* (Education Resources Information Center), *Higher Education Abstracts* (Claremont Graduate University), *Multicultural Education Abstracts* (T&F), *Sociology of Education Abstracts* (T&F).

Microfilm copies of issues and chapters are available in 16mm and 35mm, as well as microfiche in 105mm, through University Microfilms, Inc., 300 North Zeeb Road, Ann Arbor, Michigan 48106-1346.

www.josseybass.com

THE ASSOCIATION FOR INSTITUTIONAL RESEARCH was created in 1966 to benefit, assist, and advance research leading to improved understanding, planning, and operation of institutions of higher education. Publication policy is set by its Publications Committee.

For information about the Association for Institutional Research, write to the following address:

AIR Executive Office
1435 E. Piedmont Drive
Suite 211
Tallahassee, FL 32308-7955

(850) 385-4155

air@mailer.fsu.edu
http://airweb.org

CONTENTS

EDITOR'S NOTES

Attracting and retaining women in science, technology, engineering, and mathematics (STEM) fields at all levels continues to be a major issue of national concern (Goan, Cunningham, and Carroll, 2006). Although women have increased their presence in STEM fields over the past five decades (Hill, Corbett, and St. Rose, 2010), taking a closer look at the data shows women remaining underrepresented in key areas of study that are vital to economic growth and workforce development within the United States. Moreover, it is a progressive problem in that the presence of women in STEM declines toward the highest levels of the profession (Blickenstaff, 2005). This problem is especially puzzling given that women outnumber men on most college campuses across the country and women have excelled in other areas of study such as business, law, and medicine. Thus, the question remains: Why have women not progressed in similar ways within STEM fields?

This volume takes a comprehensive look at attracting and retaining women at various levels along the STEM pipeline, from early interest and preparation in math and science at the K–12 level to studying STEM at undergraduate and graduate levels to establishing a professional career in STEM. Further, the chapters within this volume raise important questions about the future of STEM education relative to stimulating interest, increasing persistence and degree completion, and designing studies to examine the experiences of women in STEM. Each chapter concludes with recommendations for institutional research, policy, and practice.

The first four chapters in the volume focus on women in STEM at the undergraduate level. In the first chapter, Casey A. Shapiro and Linda J. Sax discuss the literature relative to women's decision to select a STEM major—a process that starts as early as middle school for most students. They conclude the chapter with recommendations for institutional research and future study on gender issues in STEM. The second chapter, by Joy Gaston Gayles and Frim D. Ampaw, focuses on the influence of the college experience on degree completion in STEM. In particular, the authors discuss the importance of examining differential effects of academic and social experiences on bachelor's degree completion in STEM. Many colleges and universities have created living-learning communities to increase the retention of women in STEM. Chapter Three, by Karen Kurotsuchi Inkelas, gives an overview of the roles and functions of living-learning communities and summarizes empirical evidence on their effectiveness for increasing persistence and retention of women in STEM majors. In the fourth chapter, Dimitra Lynette Jackson and Frankie Santos Laanan discuss the major role that community colleges can play in educating the next generation of women in STEM. Jackson and Lanaan share the results from a national study on the experiences of women in STEM who transferred from the community college system, and they offer recommendations for future research.

Chapters Five and Six of this volume focus on women in STEM beyond the undergraduate years. In Chapter Five, Darnell Cole and Araceli Espinoza use social cognitive career theory to examine the postbaccalaureate career goals for women in STEM. On the basis of their findings, the authors offer recommendations for future research on the connection among self-efficacy, the college experience, and career goals for women in STEM. Frim D. Ampaw and Audrey J. Jaeger in Chapter Six discuss the experiences of women doctoral students in the sciences and factors that have an impact on doctoral degree completion. A major finding from their study was the importance of assistantship opportunities for women in STEM at the graduate level. The authors conclude the chapter with recommendations for research and practice on retaining women in STEM at the graduate level.

The final two chapters in the volume have specific implications for future research and research design concerning women in STEM. Dawn R. Johnson raises a critical issue in Chapter Seven, the problem of analyzing the experiences of women in STEM in the aggregate. Treating women as a homogeneous group obscures important racial and ethnic differences among the experiences of women in STEM. The chapter concludes by discussing the importance of addressing intersecting identities among women in STEM, which has implications for future research designs. The final chapter, by Sylvia C. Nassar-McMillan, Mary Wyer, Maria Oliver-Hoyo, and Jennifer Schneider, presents results from an NSF-funded project on scale development for two new research tools designed to measure students' perceptions and attitudes about science and scientists. The authors offer recommendations for how these tools can be used in future research and in practice.

Joy Gaston Gayles
Editor

References

Blickenstaff, J. C. "Women and Science Careers: Leaky Pipeline or Gender Filter?" *Gender and Education*, 2005, 17(4), 369–386.

Goan, S., Cunningham, A., and Carroll, C. *Degree Completions in Areas of National Need, 1996–97 and 2001–02.* Washington, DC: National Center for Education Statistics, 2006. Retrieved October 7, 2011, from http://nces.ed.gov/Pubsearch/pubsinfo. asp?pubid=2006154.

NEW DIRECTIONS FOR INSTITUTIONAL RESEARCH • DOI: 10.1002/ir

Hill, C., Corbett, C., and St. Rose, A. *Why So Few? Women in Science, Technology, Engineering, and Mathematics.* Washington, DC: American Association of University Women, 2010.

JOY GASTON GAYLES is an associate professor of higher education at North Carolina State University. Her research examines how college has an impact on student learning and development, most notably for student athletes and students in STEM.

NEW DIRECTIONS FOR INSTITUTIONAL RESEARCH • DOI: 10.1002/ir

The authors summarize common factors that contribute to women's selection of and persistence in STEM majors. The chapter concludes with a consideration of the utility of this information for institutional researchers and how they might further study issues of gender and STEM on their own campuses.

Major Selection and Persistence for Women in STEM

Casey A. Shapiro, Linda J. Sax

The U.S. federal government identifies many science, technology, engineering, and math (STEM) majors as "areas of national need" that are "crucial to national innovation, competitiveness, and well-being and in which not enough students complete degrees" (Goan, Cunningham, and Carroll, 2006, p. 1). Even with the national push to recruit and retain students in STEM, too few students enter college with the desire to pursue STEM majors, and even fewer complete STEM degrees. According to the Higher Education Research Institute (HERI) national figures on students' intended major in 2009, only 24 percent of students entering college report an interest in majoring in STEM (Pryor and others, 2009). Moreover, we can expect about half of these students to actually earn a STEM degree, as research has shown that 50 percent or more of the students who enter college with STEM career aspirations either switch to a non-STEM field or leave postsecondary education altogether (Chen and Weko, 2009; Sax, 1994b).

Women seem to be affected by these trends even more than their male counterparts. Although women comprise approximately 57 percent of college students nationwide (King, 2010), their growing presence in higher education has not translated consistently to representation within STEM fields. Among first-year college students nationwide, only 17.3 percent of women report planning to major in a STEM field, compared to 32.2 percent among men (Pryor and others, 2009). Gender disparities are especially pronounced in the fields of engineering, physical sciences, and computer sciences, all of which are designated by the U.S. government as areas of need for national innovation, competitiveness, and well-being (Goan, Cunningham, and Carroll, 2006). Further, the gender gap for students entering STEM mirrors the gap in actual STEM degree attainment five years after entering college (Huang, Taddese, and Walter, 2000).

Underrepresentation of women in STEM in the United States has economic consequences, both individually and nationally. For example, given the ongoing connection between individuals' technological skills and their economic opportunities (Bystydzienski and Bird, 2006), women's economic independence may be hindered by underparticipating in the technological industries of the twenty-first century (Weinman and Cain, 1999). Further, in light of the national call to action for research in science and technology, women's underrepresentation in STEM signifies a loss of potential talent and innovation that may have an impact on the ability of the United States to remain globally competitive in science and engineering.

Furthermore, the technological and scientific professional workforce stands to benefit from diverse perspectives. In 2005, women constituted 47 percent of the college-educated U.S. labor force, but only 27 percent of the entire science and engineering workforce (National Science Board, 2008). Creating opportunities for more women to enter and be successful in STEM fields will contribute to diversifying STEM perspectives, ultimately making scientific research more vigorous and complete (Blickenstaff, 2005).

Given that women's underrepresentation in the STEM workforce is largely rooted in their selection of college major, this chapter summarizes the factors that contribute to women's selection of and persistence in STEM majors. We begin with a review of how the educational context shapes women's interest in STEM, and then we move to an overview of the major social and cultural influences beyond the classroom. The chapter concludes with a discussion of the utility of this information for institutional researchers and how they might further study issues of gender and STEM on their own campuses.

The Role of Educational Settings in Women's Pursuit of STEM

A significant portion of the literature examining factors related to women's decision to major in STEM focuses on formal and informal educational settings. In particular, women's experiences with the curriculum as well as their academic interactions with instructors and peers are influential in shaping women's interest in and longer-term commitment to STEM (Astin and Sax, 1996; Kinzie, 2007; Margolis, Fisher, and Miller, 2000; Sax, 1994a, 1994b, 2001; Seymour and Hewitt, 1997). Thus this section focuses on four commonly identified predictors of women's interest in and decision to enter a STEM major: (1) middle and high school preparation and curriculum, (2) culture and pedagogy in college-level STEM, (3) interactions with teachers and faculty, and (4) peer-and-curriculum connections.

Middle and High School Preparation and Curriculum. Preparation in science and mathematics during the middle and high school years is often cited as an important influence on women's decisions to enter or exit STEM majors in college. In fact, Astin and Sax (1996) note that past academic achievement and ability are the "most consistent and important predictors of students' interest in science" (p. 106).

Critical junctures for women's interest and participation in STEM occur as early as middle school. Kinzie (2007) found that in eighth grade women's pathways into or out of STEM are already beginning to be formed. Kinzie suggests that math achievement is particularly important, as low math achievement in eighth grade may play a role in the number of math courses women take in high school, thereby limiting the number of women who have adequate preparation to enter college-level math and science courses.

Differential experiences in STEM continue for women and men at the high school level. In their qualitative study of more than 450 students from seven institutions, Seymour and Hewitt (1997) found that inadequate high school preparation in science and mathematics was an important factor for those who left science, mathematics, and engineering majors. Interestingly, a higher percentage of women reported concerns with inadequate high school preparation than did men (Seymour and Hewitt, 1997).

Inadequate early preparation is also problematic for women in computer science (Margolis, Fisher, and Miller, 2000). Margolis and her colleagues found that at Carnegie Mellon, a university with a highly selective computer science programming department, 40 percent of first-year men majoring in computer science reported having passed the Advanced Placement computer science test while in high school, which allowed them to "pass out" or "test out" of an introductory-level programming class in college. However, none of the first-year women had placed out of this introductory-level programming class. Thus, among first-year computer science majors, men reported having more advanced knowledge of the field than women, which can negatively influence women's confidence in their abilities to succeed in that field.

As demonstrated by these and many other studies, women's differential STEM course taking and preparation in the precollege years can limit their ability to access STEM careers later on. Women who do not take the math and science courses needed to access a career in science, math, and engineering are often unable to stay in the science/mathematics pipeline (Huang and Brainard, 2001).

Culture and Pedagogy in College-Level STEM. The unique culture of undergraduate science courses has been identified as a factor influencing STEM major selection and success. In particular, the competitive nature often fostered in STEM courses is an aspect of science pedagogy that can deter women from selecting or remaining in a STEM major (Seymour and Hewitt, 1997; Strenta and others, 1994). Seymour and Hewitt (1997) suggest that "women do not find competition a meaningful way to receive feedback" (p. 263) and may even find it to be offensive.

What aspects of science pedagogy cause the environment to feel too competitive for many women? First, STEM courses are often taught in a large, lecture format, which adds to a fierce competition for grades and weeding out of students from STEM majors (Seymour and Hewitt, 1997). Astin and Sax (1996) also note that faculty in the sciences are more likely to grade on a curve, which promotes competition among students. Additionally, the weed-out system discourages collaborative work, instead reinforcing the notion that individuals should take responsibility only for their own learning (Seymour and Hewitt, 1997).

Strenta and others (1994) also found that regardless of gender, students described science and engineering professors as relatively unresponsive, not dedicated, and not motivating compared to faculty in the humanities and social sciences. The classroom climate can disproportionately affect women in STEM, producing feelings of depression about their work and lower self-confidence among STEM women compared to women in the humanities (Strenta and others, 1994). More recent evidence also indicates that the sciences have yet to fully embrace more collaborative teaching styles. Nelson Laird, Garver, and Niskodé (2007) found that faculty in the hard disciplines (for example, agriculture, biology, chemistry, and engineering) spend on average 16 percent more time lecturing and significantly less time using active learning practices compared to the faculty in soft disciplines (e.g., communications, education, English, and sociology).

For women, having access to real-world applications of science may be particularly important to reinforcing their decision to major in STEM (Hyde and Gess-Newsome, 2000; Margolis, Fisher, and Miller, 2000). Whether in the classroom or in a lab setting, when faculty use practical and active teaching strategies, women report gains in learning and confidence (Hyde and Gess-Newsome, 2000). These teaching techniques have been found to help women connect theory to practical real-life situations, which increases their self-reported learning and confidence in their abilities to succeed in their chosen STEM field (Hyde and Gess-Newsome, 2000).

Margolis, Fisher, and Miller (2000) also stress the importance of connecting computer science to the real world and applying it to social contexts in order to retain women in computer science. Indeed, the opportunity to improve the social world could be "a powerful incentive" to study STEM (Miller, Rosser, Benigno, and Zieseniss, 2000, p. 140). However, women often do not perceive STEM as a vehicle for improving the human condition, a perception that discourages many of them from persisting in STEM fields (Sax, 1994b, 2001).

Interactions with Teachers and Faculty. In addition to the pedagogy used in many STEM courses, faculty also can have an impact on women's interest and retention in STEM majors through their classroom interactions. Ideally, such interactions will promote student interest in STEM, especially if students view faculty as role models for their own future STEM careers. Indeed, students who encounter role models within the scientific community are more likely to follow up on their initial science aspirations (Astin and Sax, 1996). Research also suggests that female role models have more positive effects on women's math performance than do male role models (Marx and Roman, 2002).

Unfortunately, because women faculty are underrepresented in STEM, female students have limited access to same-sex role models and mentors compared to men. As Blickenstaff (2005) explains, "a low proportion of women in a discipline probably sends a message to girls that the discipline is unattractive to women, and they should avoid it too" (p. 376). One way to combat this perception is to expose female students to women who are successful in STEM careers while also successful at combining work and family responsibilities (Kahveci, Southerland, and Gilmer, 2007; Kim, Fann, and Misa-Escalante, 2009). Faculty and professional role models can help women students by bolstering their confidence and encouraging them to see themselves as successful in STEM majors and careers in the future, allowing them to overcome some of the negative stereotypes about having a career in STEM, and encouraging discussion of their own experiences and strategies for working through barriers in STEM fields (Kim, Fann, and Misa-Escalante, 2009).

Although faculty as role models can bolster women's interest in STEM, students' interactions with instructors can also have negative ramifications for their interest and retention in STEM majors. In their influential report, Hall and Sandler (1982) describe the "chilly climate" for women in college classrooms, which put women at a "significant educational disadvantage" (p. 3). The authors explain that regardless of the context (classroom, lab, and so on), some student-faculty interactions may bolster women's confidence, intellectual development, and aspirations, while others can dampen women's ambitions, which Hall and Sandler argue is especially significant for those women in areas that are traditionally viewed as masculine fields, including science and engineering.

Subtle discriminatory practices have been documented specifically within STEM majors (Seymour and Hewitt, 1997; Wasburn and Miller, 2004–05). Faculty have been described as excluding women from activities in the classroom and subjecting them to different grading practices than their male peers (Seymour and Hewitt, 1997). Wasburn and Miller

(2004–05) found that nearly one-third of female respondents felt professors in their technology classes treated female and male students unequally. Furthermore, 20 percent of the female respondents did not feel comfortable asking questions in class, and almost one-quarter did not feel comfortable going to their technology professor for assistance outside of class, which ultimately had negative implications for their persistence in technology majors. Faculty interactions can also negatively affect women's mathematical self-confidence (Sax, 1994a, 2008), discourage women from persisting in science, and make them feel unwelcome in the classroom (Seymour and Hewitt, 1997), which, as noted previously, has also been shown to influence women's decisions to leave STEM fields.

Peer-and-curriculum Connections. The peer culture cultivated in educational settings can also have implications for women's commitment to STEM. As noted earlier, the pedagogy often employed in STEM classrooms tends not to facilitate positive interactions among peers (Seymour and Hewitt, 1997; Strenta and others, 1994). More specifically, many STEM courses encourage competition for grades, which promotes an emphasis on individual success rather than on collaborative learning (Strenta and others, 1994; Astin and Sax, 1996, Seymour and Hewitt, 1997). Such an environment can be especially discouraging to women, who tend to prefer more cooperative forms of learning (Barker and Garvin-Doxas, 2004).

Peer interactions have also been shown to affect the classroom climate. In fact, Colbeck, Cabrera, and Terenzini (2001) suggest that the "chilly climate" for women in engineering results from peer interactions rather than from student-faculty interactions. Colbeck and colleagues found significant gender differences in students' perceptions of how peers interacted with male and female engineering students, whereby women were significantly more likely to report that male students treated female students differently, both generally and in collaborative learning situations. However, this differential treatment did not have a significant effect on women's academic and career outcomes. In other words, women who perceive a negative classroom climate may nevertheless remain steadfast in their commitment to STEM (Seymour and Hewitt, 1997; Wyer, 2003).

Another consideration of the peer culture from a curricular perspective is the impact that the proportion of females in STEM majors has on women's interest and retention in STEM. It might seem logical that women benefit from having more women enrolled in STEM majors; however research on this topic has yielded largely mixed results. Cohoon (2001) examined computer science departments across the state of Virginia and found that from both faculty and student perspectives the proportion of females enrolled in the major was the strongest predictor of women's attrition from computer science majors, such that computer science departments with a higher proportion of females enrolled were more likely to retain those women at a rate comparable to men. The authors conclude that the proportion of female students in the major matters, as having more women enrolled in the major provides same-sex peer support, which can encourage persistence in computer science (Margolis, Fisher, and Miller, 2000).

Sax (1996) also found that the proportion of women in one's major predicted women's persistence in the major; however, the impact of gender composition was minimal on other cognitive and affective outcomes (for example, academic and mathematical self-confidence, academic achievement, satisfaction with the major). Sax's research revealed that the effect of gender composition is typically explained by other aspects of the major (for example, level of competitiveness, faculty attributes, and precollege orientations of students enrolled in the major).

Forces Beyond the Classroom

In addition to educational environments, women's interactions and experiences outside the classroom can affect their interest and retention in STEM majors (Brainard and Carlin, 1998; Han, Sax, and Kim, 2007; Margolis, Fisher, and Miller, 2000; Seymour and Hewitt, 1997; Zhao, Carini, and Kuh, 2005). This section focuses on four key predictors of women's interest and decision to enter a STEM major: (1) self-confidence, (2) sense of belonging in the STEM culture, (3) family influences and expectations, and (4) peers and social/cocurricular connections.

Self-Confidence. Women consistently express lower levels of academic and mathematical confidence than their male peers, even when their demonstrated academic and mathematical abilities are equal to men's (Sax, 1994a, 2008). This also holds true for women's self-reported cognitive gains, such that women tend to report higher college grade point averages (GPA) compared to men yet tend to perceive fewer gains in their quantitative and analytical skills during college (Zhao, Carini, and Kuh, 2005). Thus one could argue that there is a discrepancy between women's perceived ability and their demonstrated achievement (Astin and Sax, 1996). In other words, women may not leave science and engineering majors for lack of academic ability, but rather because of a lack of scientific self-confidence (Brainard and Carlin, 1998).

Sense of Belonging in the STEM Culture. Research has consistently documented the importance of women's sense of belonging within STEM (Brainard and Carlin, 1998; Han, Sax, and Kim, 2007; Margolis, Fisher, and Miller, 2000; Seymour and Hewitt, 1997). Women are often exposed to attitudes about what it means to major or have a career in STEM, which can negatively influence their STEM-related self-concept (Astin and Sax, 1996). Students report these gendered attitudes can start early in one's academic career in the form of gender socialization either toward or away from science (Kim, Fann, and Misa-Escalante, 2009). Additionally, stereotypes of what it means to be an "engineer" continue to be primarily defined in terms of men (Tonso, 2006). It is this notion along with assumptions about the lifestyle of STEM careers that may conflict with gendered stereotypes regarding women's roles in the home, making STEM majors and careers less appealing to women (Kim, Fann, and Misa-Escalante, 2009).

In their study of computer science students, Margolis, Fisher, and Miller (2000) found that women lost confidence and interest in computer science because they felt they did not fit with the stereotypical view of a computer scientist. Piatek-Jimenez (2008) also found that women's stereotypical views about mathematical careers influenced their desire to have a career as a mathematician. In other words, women tend not to identify with traditional notions of what it means to have a career in STEM and may therefore choose not to pursue majors or careers in STEM.

Related to these influences are the explicit and implicit messages about the masculine nature of math and science. Explicit messages can include stereotypes that math is a masculine field and therefore not a discipline in which females can excel (Nosek, Banaji, and Greenwald, 2002). Women may not consciously acknowledge these masculine stereotypes of STEM, but implicitly these perceptions can influence the type of major they select (Nosek, Banaji, and Greenwald, 2002). Furthermore, when internalized by men and women, these perceptions can have an impact on women's sense of identification with STEM (Margolis, Fisher, and Miller, 2000; Nosek, Banaji, and Greenwald, 2002; Wyer, 2003).

Despite traditionally masculine views about science, it is important to note that women have made significant strides in certain areas of STEM, such as biological science, chemistry, and agricultural science (National Science Foundation, 2008). Some argue that this shift in women's participation in biological sciences is due to the nature of the field, which tends to more directly and explicitly address human problems (Miller, Rosser, Benigno, and Zieseniss, 2000).

Family Influences and Expectations. As noted previously, role models can influence women's desire to enter and be retained in STEM majors during college. Parents can also serve as role models for women interested in STEM. In fact, a woman is more likely to pursue a career in STEM if one or both of her parents had a career in these fields (Astin and Sax, 1996). For both men and women, having a father who is an engineer was associated with persistence toward a career in STEM (Sax, 1994b). Seymour and Hewitt (1997) also note the importance of having practical role models and mentors for women, including professional women in their own family.

However, parents are not always a positive influence. Historically, many parents have accepted the stereotype that men are more apt to succeed in STEM than women (Vetter, 1996). Vetter argued that in terms of their children's achievements in mathematics and science, "as a group, parents have lower educational aspirations for daughters than for sons" (p. 32). These lowered expectations can manifest in differential pressure from parents for their daughters and sons to persist in STEM, often resulting in women feeling less pressure from parents to complete a STEM major (Seymour and Hewitt, 1997).

Students' own notions about family and work can also influence their interest in majoring STEM fields (Han, Sax, and Kim, 2007). Even when men and women report equal levels of academic ability, first-year college women are still more likely to anticipate that conflicts between work and family responsibilities will be a barrier to career success (Hawks and Spade, 1998). Therefore, it seems that early in college women already perceive careers in STEM as incompatible with successfully raising a family, a perception that has implications for their desire to major in STEM in college.

Peers and Social/Cocurricular Connections. In addition to the role of peers in the academic setting (discussed earlier), the peer culture outside the classroom plays a role in women's lives and career decision making (Riegle-Crumb, Farkas, and Muller, 2006). According to Riegle-Crumb, Farkas, and Muller (2006), high school women's friendship groups influence their advanced-course-taking patterns, especially in the areas of math and science. Specifically, friendship groups that have a high combination of female friends and performance in math and science facilitate women's persistence in advanced courses, such as calculus and physics. Thus one could argue that for girls there is a relationship between high school friendship groups and the math and science preparation courses that are often needed to access a career in STEM.

Peers can also serve as role models for undergraduate women interested in STEM (Kahveci, Southerland, and Gilmer, 2007). Interactions with peers can provide women with an avenue to exchange information, find study partners, and create informal peer role models (Hyde and Gess-Newsome, 2000; Kahveci, Southerland, and Gilmer, 2007). Formal big-sister little-sister programs that connect older and younger female students in engineering also seem to help female students persist (Brainard and Carlin, 1998). However, Seymour and Hewitt (1997) found that women in STEM are often surrounded by peers who make them feel unwelcome, take them less seriously, or treat them with hostility. Margolis, Fisher, and Miller (2000) also found that interacting with male peers can often unravel women's confidence, especially when male students make comments that reinforce women's notions that they don't belong, such as "You only got into computer science because you are a girl" (p. 117). In fact, Margolis argues that "peer support is . . . vital" (p.124), as having a community of other women and sympathetic men in their major to share their struggles helps women feel they are not alone.

Women's Participation in STEM: An Agenda for Institutional Researchers

As described throughout this chapter, women's underrepresentation in STEM results from a complex array of forces stemming from education, family, and the larger society. These influences are generally outside the control of campus-based institutional researchers, but there is an opportunity, through documentation and new research, for institutional researchers to promote women's participation in STEM in college. Consider these recommendations for campus-based institutional research offices.

Document the Participation of Women in STEM. Though it is common practice for institutional research offices to track enrollment of students in STEM, even broken down by gender, it is less common to examine gender differences in *persistence* within STEM. Given that students often do not declare their major until their junior year, a more complete picture of the STEM pipeline in college may require more than just tracking students by major field. Presuming that STEM-based course-taking patterns can be identified, researchers should consider using student transcripts to analyze gender differences in course-taking patterns that lead to STEM major selection and ultimate degree attainment. Researchers should also look for gender-based patterns in the decision *not* to pursue STEM. Are there specific course-taking patterns that lead to the selection of certain non-STEM majors, and does this differ by gender?

Be Attuned to Variations Within STEM Fields. Even though gender disparities persist in historically male-dominated fields such as physical science, computer science, and engineering, women have made significant progress in certain STEM fields, such as the biological sciences and chemistry. Thus any analysis of gender differences within STEM must distinguish *among* the various STEM fields.

Assess the Climate for Women in STEM. Researchers have historically observed a "chilly climate" for women in STEM fields, and yet today women are thriving within STEM departments at numerous colleges across the country. In other words, the climate for women in STEM has not remained chilly everywhere. Institutional research or assessment offices should regularly assess the climate for all students in STEM, with a particular eye on the experiences of underrepresented students such as women and students of color. Again, keeping in mind variations *within* STEM, institutional researchers may wish to focus specifically on a number of aspects of the STEM climate:

- *Pedagogy and classroom climate.* To what extent does STEM pedagogy involve lecture versus collaborative learning? To what extent are real-world applications emphasized in instruction? To what extent do grading practices promote competition among students? What is the nature of student-faculty interaction? To what extent are students exposed to female STEM professionals and to professionals who have successfully combined scientific work and family responsibilities? When considering the STEM climate, gather a variety of perspectives, as there may be important perceptual differences between students and faculty, between women and men, and so on. Also consider the extent to which the STEM climate is distinct from other academic climates on campus.

- *Interactions with peers.* What is the nature of the academic and social interactions among peers in STEM? To what extent do peer interactions reflect competition versus collaboration? To what extent are women included in the formation of study groups? What are the gender dynamics of study groups? Are there social support groups for women in STEM?
- *Science identity.* Assess gender differences in students' interest and confidence in STEM. Regularly survey students to allow tracking STEM-related confidence over time.
- *Portrayal of science.* Assess how STEM fields are portrayed on your campus. To what extent do marketing materials and STEM departmental websites appeal to both male and female students?

Ultimately, in addition to documenting the participation of women in STEM on their campus, institutional research offices should assess the conditions for women in STEM. Through research and assessment, institutional researchers can serve as agents of change in advancing opportunities for women in STEM. Naturally, to be most effective such work requires coordination between, and collaboration among, other campus units, ranging from academic to student affairs, and directly involving faculty and students from specific STEM departments.

References

Astin, H. S., and Sax, L. J. "Developing Scientific Talent in Undergraduate Women." In C. Davis, A. B. Ginorio, C. S. Hollenshead, B. B. Lazarus, P. M. Rayman, and Associates (eds.), *The Equity Equation: Fostering the Advancement of Women in the Sciences, Mathematics, and Engineering.* San Francisco: Jossey-Bass, 1996.

Barker, L. J., and Garvin-Doxas, K. "Making Visible the Behaviors That Influence Learning Environment: A Qualitative Exploration of Computer Science Classrooms." *Computer Science Education,* 2004, *14*(1), 119–145.

Blickenstaff, J. C. "Women and Science Careers." *Gender and Education,* 2005, *17*(2), 369–386.

Brainard, S. G., and Carlin, L. "A Six-Year Longitudinal Study of Undergraduate Women in Engineering and Science." *Journal of Engineering Education,* 1998, *87*(4), 17–27.

Bystydzienski, J. M., and Bird, S. R. *Removing Barriers: Women in Academic Science, Technology, Engineering, and Mathematics.* Bloomington: Indiana University Press, 2006.

Chen, X., and Weko, T. *Students Who Study Science, Technology, Engineering, and Mathematics (STEM) in Postsecondary Education*. Washington, D.C.: National Center for Education Statistics, 2009. Retrieved October 7, 2011, from http://nces.ed.gov/pubs2009/2009161.pdf.

Cohoon, J. M. "Toward Improving Female Retention in the Computer Science Major." *Communications of the ACM*, 2001, 44(5), 108–114.

Colbeck, C. L., Cabrera, A. F., and Terenzini, P. T. "Learning Professional Confidence: Linking Teaching Practices, Students' Self-Perceptions, and Gender." *Review of Higher Education*, 2001, 24(2), 173–191.

Goan, S., Cunningham, A., and Carroll, C. *Degree Completions in Areas of National Need, 1996–97 and 2001–02*. Washington, D.C.: National Center for Education Statistics, 2006. Retrieved October 7, 2011, from http://nces.ed.gov/Pubsearch/pubsinfo.asp?pubid=2006154.

Hall, R. M., and Sandler, B. R. *The Classroom Climate: A Chilly Climate for Women?* Washington, D.C.: Association of American Colleges, 1982.

Han, J. C., Sax, L. J., and Kim, K. A. "Having the Talk: Engaging Engineering Students in Discussions on Gender and Equity." *Journal of Women and Minorities in Science and Engineering*, 2007, 13(2), 145–163.

Hawks, B. K., and Spade, J. Z. "Women and Men Engineering Students: Anticipation of Family and Work Roles." *Journal of Engineering Education*, 1998, 87(3), 249–256.

Huang, G., Taddese, N., and Walter, E. *Entry and Persistence of Women and Minorities in College Science and Engineering Education*. Washington, D.C.: National Center for Education Statistics, 2000. Retrieved October 7, 2011, from http://www.nces.ed.gov/pubsearch/pubsinfo.asp?pubid=2000601.

Huang, P. M., and Brainard, S. G. "Identifying Determinants of Academic Self-Confidence Among Science, Math, Engineering, and Technology Students." *Journal of Women and Minorities in Science and Engineering*, 2001, 7(4), 315–337.

Hyde, M. S., and Gess-Newsome, J. "Factors That Increase Persistence of Female Undergraduate Science Students." In J. Bart (ed.), *Women Succeeding in the Sciences: Theories and Practices Across Disciplines*. West Lafayette, Ind.: Purdue University Press, 2000.

Kahveci, A., Southerland, S. A., and Gilmer, P. J. "From Marginality to Legitimate Peripherality: Understanding the Essential Functions of a Women's Program." *Science Education*, 2007, 92(1), 33–64.

Kim, K., Fann, A., and Misa-Escalante, K. "Engaging Women in Computer Science and Engineering: Insights from a National Study of Undergraduate Research Experiences." Los Angeles: Center for Embedded Network Sensing, 2009. Retrieved October 6, 2011, from http://escholarship.org/uc/item/6w6295sp.

King, J. E. *Gender Equity in Higher Education: 2010*. Washington, D.C.: American Council on Education, 2010.

Kinzie, J. "Women's Paths in Science: A Critical Feminist Analysis." *New Directions for Institutional Research*, 2007, 133, 81–93.

Margolis, J., Fisher, A. and Miller, F. "The Anatomy of Interest: Women in Undergraduate Computer Science." *Women's Studies Quarterly*, 2000, 28(1/2), 104–127.

Marx, D. M., and Roman, J. S. "Female Role Models: Protecting Women's Math Test Performance." *Personality and Social Psychology Bulletin*, 2002, 28(9), 1183–1193.

Miller, P. H., Rosser, S. V., Benigno, J. P., and Zieseniss, M. L. "A Desire to Help Others: Goals of High-Achieving Female Science Undergraduates." *Women's Studies Quarterly*, 2000, 28(1/2), 128–142.

National Science Board. *Science and Engineering Indicators 2008*. Arlington, Va.: National Science Foundation (vol. 1, NSB 08-01; vol. 2, NSB 08-01A), 2008. Retrieved October 7, 2011, from http://www.nsf.gov/statistics/seind08/toc.htm.

NEW DIRECTIONS FOR INSTITUTIONAL RESEARCH • DOI: 10.1002/ir

National Science Foundation, Division of Science Resources Statistics. *Science and Engineering Degrees: 1966–2006* (Detailed Statistical Tables NSF 08-321). Arlington, Va.: National Science Foundation, 2008. Retrieved October 7, 2011, from http://www.nsf.gov/statistics/nsf08321/pdf/nsf08321.pdf.

Nelson Laird, T. F., Garver, A. K., and Niskodé, A. S. "Gender Gaps: Understanding Teaching Style Differences Between Men and Women." Paper presentation at the Annual Forum of the Association for Institutional Research, Kansas City, Mo., June 2–6, 2007.

Nosek, B. A., Banaji, M. R., and Greenwald, A. G. "Math = Male, Me = Female, Therefore Math ≠ Me." *Journal of Personality and Social Psychology*, 2002, 83(1), 44–59.

Piatek-Jimenez, K. "Images of Mathematicians: A New Perspective on the Shortage of Women in Mathematical Careers." *ZDM—The International Journal on Mathematics Education*, 2008, 40(4), 633–646.

Pryor, J. H., and others. *The American Freshman: National Norms for Fall 2009, Expanded Edition.* Los Angeles: Higher Education Research Institute, UCLA, 2009.

Riegle-Crumb, C., Farkas, G., and Muller, C. "The Role of Gender and Friendship in Advanced Course Taking." *Sociology of Education*, 2006, 79(3), 206–228.

Sax, L. J. "Mathematical Self-concept: How College Reinforces the Gender Gap." *Research in Higher Education*, 1994a, 35(2), 141–166.

Sax, L. J. "Retaining Tomorrow's Scientists: Exploring the Factors That Keep Male and Female College Students Interested in Science Careers." *Journal of Women and Minorities in Science and Engineering*, 1994b, 1(1), 45–61.

Sax, L. J. "The Dynamics of Tokenism: How College Students Are Affected by the Proportion of Women in Their Major." *Research in Higher Education*, 1996, 37(4), 389–425.

Sax, L. J. "Undergraduate Science Majors: Gender Differences in Who Goes to Graduate School." *Review of Higher Education*, 2001, 24(2), 153–172.

Sax, L. J. *The Gender Gap in College: Maximizing the Development Potential of Women and Men.* San Francisco: Jossey-Bass, 2008.

Seymour, E., and Hewitt, N. *Talking About Leaving: Why Undergraduates Leave the Sciences.* Boulder, Colo.: Westview Press, 1997.

Strenta, A. C., and others. "Choosing and Leaving Science in Highly Selective Institutions." *Research in Higher Education*, 1994, 35(5), 513–547.

Tonso, K. L. "Student Engineers and Engineer Identity: Campus Engineer Identities as Figured World." *Cultural Studies of Science Education*, 2006, 1(2), 237–307.

Vetter, B. M. "Myths and Realities of Women's Progress in the Sciences, Mathematics, and Engineering." In C. Davis, A. B. Ginorio, C. S. Hollenshead, B. B. Lazarus, P. M. Rayman, and Associates (eds.), *The Equity Equation: Fostering the Advancement of Women in the Sciences, Mathematics, and Engineering.* San Francisco: Jossey-Bass, 1996.

Wasburn, M. H., and Miller, S. G. "Retaining Undergraduate Women in Science, Engineering, and Technology: A Survey of a Student Organization." *Journal of College Student Retention*, 2004–05, 6(2), 155–168.

Weinman, J. and Cain, L. "Technology—The New Gender Gap." *Technos: Quarterly for Education and Technology*, 1999, 8(1), 9–12.

Wyer, M. "Intending to Stay: Images of Scientists, Attitudes Toward Women, and Gender as Influences on Persistence Among Science and Engineering Majors." *Journal of Women and Minorities in Science and Engineering*, 2003, 9(1), 1–16.

Zhao, C. M., Carini, R. M., and Kuh, G. D. "Searching for the Peach Blossom Shangri-La: Student Engagement of Men and Women SMET Majors." *Review of Higher Education*, 2005, 28(4), 503–525.

CASEY A. SHAPIRO *is a postdoctoral scholar at UCLA's Center for Educational Assessment. Dr. Shapiro's research focuses on gender and undergraduate science education, with a focus on the influence of friendship groups on women's pursuit of STEM fields.*

LINDA J. SAX *is a professor of higher education in the Graduate School of Education and Information Studies at UCLA. Dr. Sax's research focuses on gender differences in college student development, with an emphasis on women in STEM fields. She is the author of* The Gender Gap in College: Maximizing the Developmental Potential of Women and Men *and is the principal investigator on a nationwide study of long-term shifts in the gender gap in STEM fields, funded by the National Science Foundation.*

2

This chapter summarizes the results from a study that examined differential effects of the college experience on degree completion in STEM by gender. The finding that some experiences are more or less influential for women in the study has important implications for institutional research.

Gender Matters: An Examination of Differential Effects of the College Experience on Degree Attainment in STEM

Joy Gaston Gayles, Frim D. Ampaw

Although more women than men are enrolled in college within the United States, women remain underrepresented in critical areas of study such as science, technology, engineering, and mathematics (STEM). This is particularly concerning given that STEM fields of study are vital to the economic growth and workforce development within the United States (Commission on Professionals in Science and Technology, 2006; National Science Board, 2006). In order for the United States to maintain its status as a competitor in addressing global issues, it will be imperative to increase and diversify the U.S. STEM workforce. The loss of women in STEM fields at critical junctures of the education pipeline has received national attention over the past few decades, and federal support has been earmarked for attracting and retaining women and other underrepresented populations in STEM fields. Although it is true that women have made gains in bachelor's degree attainment in STEM over the past thirty years, a closer look at the data shows that they have not reached parity with their male counterparts. A critical issue of concern is the high number of women who enter college with an interest in STEM and the low number of women who actually complete a STEM bachelor's degree six years later (Huang, Taddese, and Walter, 2000).

Literature on Women in STEM

There is a vast amount of empirical literature on the status and experiences of women in STEM in response to the national attention directed toward

NEW DIRECTIONS FOR INSTITUTIONAL RESEARCH, no. 152, Winter 2011 © Wiley Periodicals, Inc. Published online in Wiley Online Library (wileyonlinelibrary.com) • DOI: 10.1002/ir.405

increasing the representation of women in these critical areas. From the extant literature, we know that student background characteristics and institutional factors affect the persistence of women in STEM fields (Huang, Taddese, and Walter, 2000; Sax, 2001; Vogt, Hocevar, and Hagedorn, 2007). More specifically, research supports that high school performance in math and science is a strong indicator of success in STEM Fields (Huang, Taddese, and Walter, 2000). Parental education and income are also important factors that influence persistence in STEM fields.

Within the extant literature there is tremendous support for the positive impact of academic and social experiences on student learning and development (see Pascarella and Terenzini, 2005 for a full discussion of how college affects students). Further, women are typically found to be as engaged in their educational experiences compared to men (Sax, 1994; Zhao, Carini, and Kuh, 2005). Although these studies focus on all students, research on students in STEM fields supports that they are generally satisfied with their college experience. Generally, the college impact literature examines students in the aggregate. However, doing so can mask important gender differences concerning how college differentially affects degree completion for men and women in the STEM fields. Future research on how college has an impact on outcomes of undergraduate education for female students in these fields must address this void in the extant literature. This study contributes in that direction.

Conceptual Framework

Theories of college impact are useful in helping us understand and frame how the college experience influences various outcomes of undergraduate education, among them persistence, academic performance, and degree attainment. Some of the foundational models of college impact are Astin's Input-Environment-Outcome model (I-E-O; 1993), Tinto's theory of student departure (1993), Weidman's model of undergraduate socialization (1989), and Pascarella's general causal model (1985). Each of these models has a unique approach to examining college impact on cognitive and affective outcomes of education, but they also have commonalities that, taken together, provide a powerful framework for studying college impact more generally. First, important to understanding how college experiences influence outcomes are students' background characteristics and personal attributes. Such experiences are linked to both what students do in college and the outcome of interest. Background characteristics and precollege attributes include variables such as demographics, high school experiences, and a pre-measure of the outcome variable.

Another commonality across college impact models is the college experience as well as characteristics of the institution. Astin's theory of involvement (1984) suggests that the amount of physical and psychological energy that students invest in educational activities is positively associated with student learning and personal development. Involvement consists of activities such as engagement with faculty and participation in student groups and organizations. In Tinto's model of student departure (1993), involvement is represented by two major factors: academic and social integration. Academic and social integration both refer to the extent to which students internalize, respectively, the academic and social shared norms and values of the institution. This often occurs through significant interaction with key socialization agents such as peers and faculty. Positive integration experiences lead to the formation of deeper commitment to the institution, which increases the likelihood of persistence. Negative integration experiences weaken an individual's commitment to the institution and are linked to departure.

Many of the empirical studies on college impact make the assumption that the effects of the college experience on desired undergraduate educational outcomes are the same for all students because few studies have taken into consideration whether or not the academic and social effects of the college experience depend on gender or another demographic characteristic (Pascarella and Terenzini, 2005; Sax, 2008). In other words, there is a need for empirical studies that examine whether or not interacting with faculty, for example, has the same level of effect on academic success for both men and women, especially for those majoring in the STEM fields.

Description of the Study

This chapter reports the major findings from a study that examined the extent to which the effects of college experiences on degree completion in STEM at four-year colleges and universities differed for men and women. As stated earlier, one of the major shortcomings of the college impact literature is the inherent assumption that the effects of college are similar for all subpopulations of college students. Very few studies have been designed to examine the extent to which the effects of the college experience, such as participating in student groups and organizations, vary across factors such as race and ethnicity or gender. Thus, our research questions particularly focused on what college experiences matter in relation to six-year degree attainment in STEM and whether or not the effects of college on degree attainment are conditional on the basis of gender.

Data for this study came from a national dataset: the 1995/2001 Beginning Postsecondary Students (BPS) longitudinal study. The analytic sample consisted of 1,488 students who entered four-year colleges in 1995 and declared a science, technology, engineering, or mathematics major and were still majoring in STEM during the last observable year of enrollment. For the purpose of this study, STEM majors did not include psychology and social science majors. Although these majors are defined as sciences by the Department of Education and the National Science foundation, this study excluded them from the STEM definition since women are not considered to be underrepresented in these areas.

The variables used for this study were derived from our conceptual framework. The dependent variable was bachelor's degree attainment in the STEM major within six years at a four-year institution. The explanatory variables were divided into two categories: (1) student background characteristics and (2) college experiences. Student background characteristics consisted of measures of a student's age, gender, race or ethnicity, and socioeconomic status in the estimation. The college experiences variables consisted of institutional characteristics, academic experiences, and social experiences.

Summary of Major Findings

The results from our study yielded a number of interesting findings about which college experiences influence timely degree completion in STEM fields. We discuss the results of the first model, which examines degree completion for all students, and then the second model, where gender interacts with our variables of interest. Similar to previous research reports on women in STEM, we found that although women and men did not differ much in terms of background characteristics and precollege attributes, women were less likely to complete a degree in STEM within six years.

Results from the First Model (No Interaction Terms). The results of the logistic regression analysis indicated key factors that increased or decreased the likelihood of degree completion in STEM. Similar to previous studies and as mentioned in the previous chapter, math and science preparation at the high school level was a positive predictor of STEM degree completion. More specifically, the results showed that high school science grades had the strongest effect on degree attainment. Other background and institutional characteristics related to degree completion in STEM are parental education, family income, dependent status, and institutional type and selectivity. Students who had parents with higher levels of income and education as well as students who held dependent status were most likely to graduate with a degree in STEM within six years. Students who attended very selective institutions and students at liberal arts colleges were more likely to complete a STEM degree within six years.

Understanding how college experiences, particularly academic and social integration, influence degree attainment for students in STEM was of major interest. For the full sample, interacting with faculty had a positive effect on degree completion—particularly talking to faculty outside of the classroom, which had the largest effect on degree attainment. Surprisingly, meeting with academic advisors about course plans did not have a positive effect on degree attainment for students in STEM. The results related to attending study groups outside of class were interesting in that doing this activity sometimes had no effect on degree attainment, but engaging in this activity often had a negative impact of degree attainment as compared with students who did not attend study groups at all.

As expected, grade point average had a positive influence on degree attainment. In fact, the chances of graduating in 6 years with a degree in STEM increased 30 percentage points for every one unit increase in grade point average. Social integration, which consisted of activities such as participation in social and intramural clubs, had a small positive impact on degree attainment. Satisfaction with the campus climate had a small negative impact on degree attainment. In addition, working more than fifteen hours per week decreased the likelihood of degree attainment. However, attending school full-time increased the likelihood of degree attainment in STEM.

Results from the Second Model (Interaction with Gender). In addition to the findings concerning all students in STEM, the results from this study shed light on which factors are more or less influential by gender. Because women are so severely underrepresented in STEM majors, there is a need to better understand how college experiences influence degree completion by gender specifically. Thus we examined conditional effects of the college experience on degree attainment by gender in our analysis. The evidence from the analyses that included interaction terms suggests that the effects of college on degree attainment are not the same for men and women.

Talking to faculty outside the classroom at least sometimes had a stronger effect on degree completion for females in the study compared to males. For males to benefit from interacting with faculty, they had to engage in this activity often. In addition, it was more beneficial for women to attend school full-time compared to men.

Some of the negative effects of college on degree completion were even more severe for women in the study. For example, the negative effects of meeting with an advisor, having social contact with advisors and faculty, and attending study groups often were more dramatic for women in the study in comparison to men.

Discussion and Implications for Institutional Research

The results from this study have several implications for institutional research. At a basic level, they show that factors such as academic performance in math and science at the high school level continue to be important determinants for future success in STEM majors. The results also highlight the kinds of experiences that are meaningful to degree attainment in STEM. It is not surprising that interacting with faculty outside the classroom often was significant. Such interactions are an important part of the socialization process, when critical information is shared and discussed about the discipline that is not likely to occur as a part of classroom discussions. Institutional researchers on college campuses should try to gain a better understanding of the types of interactions that are meaningful for women in STEM. Thus, future research should seek to further specify and examine the influence of faculty-student interactions; we could not determine from our study if students in STEM were referring to interacting with faculty inside or outside of their major.

Future research might also identify and examine the influence of other types of interactions that could be meaningful such as participation in undergraduate research with a faculty member or presenting at a national conference with a faculty member. In essence, institutional researchers can benefit from identifying more specifically the kinds of interactions that occur between faculty and students outside the classroom that are most meaningful for students in STEM on their campuses.

In addition, the results from our study underscore the importance of examining differential effects of the college experience on degree completion by gender and have important implications for institutional research. When studying what works for students relative to desired outcomes of undergraduate education, researchers must not assume that the effects are similar for all students. Thus research designs should incorporate methods to examine conditional effects across gender, race, and ethnicity, etc. The findings that some types of experiences are more or less beneficial for women in STEM suggest the need to examine more closely the cultural climate within STEM fields. Such studies have the potential to help us better understand the extent to which women benefit from academic and social experiences compared to males in STEM majors, which can have important implications for policy and practice.

References

Astin, A. W. "Student Involvement: A Developmental Theory for Higher Education." *Journal of College Student Personnel*, 1984, 25, 297–308.

Astin, A. W. *What Matters in College? Four Critical Years Revisited*. San Francisco: Jossey-Bass, 1993.

Commission on Professionals in Science and Technology. *Four Decades of STEM Degrees, 1996–2004: "The Devil Is in the Details."* Washington, D.C.: Commission on Professionals in Science and Technology, 2006.

Huang, G., Taddese, N., and Walter, E. *Entry and Persistence of Women and Minorities in College Science and Engineering Education. NCES 2000601.* Washington, D.C.: National Center for Educational Statistics, 2000.

National Science Board. *Science and Engineering Indicators 2006*. Arlington, Va.: National Science Board, 2006.

Pascarella, E. T. "College Environmental Influences on Learning and Cognitive Development: A Critical Review and Synthesis." In J. Smart (ed.), *Higher Education: Handbook of Theory and Research* (Vol. 1). New York: Agathon, 1985.

Pascarella, E. T., and Terenzini, P. T. *How College Affects Students* (2nd ed.). San Francisco: Jossey-Bass, 2005.

Sax, L. J. "Mathematical Self-Concept: How College Reinforces the Gender Gap." *Research in Higher Education*, 1994, 35(2), 141–166.

Sax, L. J. "Undergraduate Science Majors: Gender Differences in Who Goes to Graduate School." *Review of Higher Education*, 2001, 24(2), 153–172.

Sax, L. J. *The Gender Gap in College: Maximizing the Developmental Potential of Women and Men.* San Francisco: Jossey-Bass, 2008.

Tinto, V. *Leaving College: Rethinking the Causes and Cures of Student Attrition* (2nd ed.). Chicago: University of Chicago, 1993.

Vogt, C., Hocevar, D., and Hagedorn, L. "A Social Cognitive Construct Validation: Determining Women and Men's Success in Engineering Programs." *Journal of Higher Education*, 2007, 78(3), 336–364.

Weidman, J. C. "Undergraduate Socialization: A Conceptual Approach." In J. C. Smart (ed.), *Higher Education: Handbook of Theory and Research* (Vol. 5). New York: Agathon, 1989.

Zhao, C., Carini, C. M., and Kuh, G. D. "Searching for the Peach Blossom Shangri-La: Student Engagement of Men and Women SMET Majors." *Review of Higher Education*, 2005, 28(4), 503–525.

JOY GASTON GAYLES is an associate professor of higher education at North Carolina State University. Her research examines how college has an impact on student learning and development, most notably for student athletes and students in STEM.

FRIM D. AMPAW is an assistant professor of educational leadership at Central Michigan University. Her research investigates the retention of doctoral, female, and minority students in the STEM fields.

This chapter provides an overview of the role and functions of living-learning programs on college campuses and summarizes empirical evidence on the effectiveness of living-learning programs in attracting and retaining women in STEM majors.

Living-Learning Programs for Women in STEM

Karen Kurotsuchi Inkelas

The underrepresentation of women in the science, technology, engineering, and mathematics (STEM) educational pipeline, as well as critical junctures that negatively affect women in their pursuit of STEM degrees and careers, has been well documented and discussed in this volume's previous chapters. This chapter takes a slightly different approach and focuses specifically on one type of undergraduate intervention and its contribution to facilitating success for women in STEM: the living-learning program (LLP). Living-learning programs are a type of learning community best described as curricular linkages designed to provide students with a deeper and more integrated learning experience (Gabelnick, MacGregor, Matthews, and Smith, 1990; Shapiro and Levine, 1999). However, living-learning programs, unlike their broader learning community counterparts, are based in college residence halls and co-join students' living environments with a particular academic theme or topic. Although LLPs as a concept were not created expressly for women in STEM, there are specific LLPs that cater to women in STEM, or STEM disciplines in general, and there is limited evidence that LLPs—respective or irrespective of theme—may be successful in fostering women's success in STEM. In this chapter, I first summarize the roles and functions of LLPs, as well as the variety of LLPs that focus on STEM-based topics, and then review the empirical evidence on LLP effectiveness and STEM success for women. I conclude with implications from this line of research for institutional research, especially when considering women in STEM.

NEW DIRECTIONS FOR INSTITUTIONAL RESEARCH, no. 152, Winter 2011 © Wiley Periodicals, Inc.
Published online in Wiley Online Library (wileyonlinelibrary.com) • DOI: 10.1002/ir.406

Roles and Functions of Living-Learning Programs

Living-learning programs have been described by a variety of phrases, among them residential learning communities, living-learning communities, living-learning centers, theme houses, and residential colleges (Inkelas and Soldner, 2011). However, all of these terms (which will hereafter be referred to as "living-learning programs" or "LLPs," in this chapter) share some common core characteristics: they are residence-hall-based programs in which students live together in the same dwelling; they are designed to integrate students' in-class and out-of-class experiences; and they seek to create a more intimate learning environment, particularly for students who attend a large university (Inkelas and Soldner, 2011).

Yet the modern living-learning program can trace its roots back to an earlier era in American higher education. Although some (for example, Chaddock, 2008) may assert that Pythagoras conducted the first LLP in 6 B.C., most observers begin the history of LLPs in the nineteenth century. Early American colonial colleges based much of their educational enterprise on British models, particularly Oxford and Cambridge Universities (Alexander, 1998). The British model was largely a residential college design, where students—alongside their instructors—lived, attended class, studied, dined, socialized, prayed, and slept in the same building (Ryan, 1992). This residential college model is not unlike contemporary residential college models with faculty and students living together and sharing a strong focus on the liberal arts.

However, by the late 1800s the British residential college model of higher education had slowly given way to the Germanic model, with its greater focus on academic disciplines and professionalization of research inquiry (Veysey, 1965; Rudolph, 1990; Ryan, 1992). The Germanic model of higher education shifted the focus of learning from a communal atmosphere to one where students specialized in a particular vocation, or major. In this educational system, the college residence became separated from academics, and thus the original spirit of the residential college was greatly diminished if not discarded (Veysey, 1965; Rudolph, 1990).

However, not everyone was enthralled with the Germanic model of higher education. Some, like Alexander Meiklejohn, philosopher and academic leader in the early twentieth century, felt that the increasingly narrow focus on academic disciplines was fragmenting the lifeblood of universities. As an alternative, Meiklejohn founded the Experimental College at the University of Wisconsin in 1927, which is widely considered to be the first modern living-learning program in the twentieth century. Using a curriculum focusing on democracy, the Experimental College integrated students' curricular, cocurricular, and living experiences and incorporated new pedagogical techniques such as team teaching and linked courses (Meiklejohn, [1932] 2001). Yet the Experimental College lasted only five years, and Meiklejohn himself noted the difficulty in maintaining such a vanguard program in the face of the dominant higher education model of the day.

Living-learning programs, though, have made a comeback during the past three decades, largely in reaction to several clarion calls for reform in American higher education. First, critics (for example, Bloom, 1987; Ravitch and Finn, 1988; Sykes, 1988) questioned the quality of undergraduate education in the late 1980s, including complaints that the curriculum was fragmented (eerily echoing Meiklejohn sixty years prior) and inexperienced; graduate student instructors were teaching to overenrolled classes, and professors were proselytizing during their lectures. Second, in response to these critiques, federal policy makers also scrutinized undergraduate education at large universities and challenged colleges and universities to redouble their efforts to improve the quality of undergraduate education by using more active and collaborative learning techniques, especially in the first year (Boyer Commission, 1998; Association of American Colleges and Universities, 2002). Third, the first-year experience became a focal point in higher education as more and more diverse populations enrolled in college but subsequently withdrew because of difficulties in making the transition to college.

Living-learning programs attempted to help ameliorate such criticisms of undergraduate education, and postsecondary institutions promoted LLPs as the "answer" to many of higher education's ills. First, LLPs are thought to provide a more intimate learning atmosphere on an otherwise vast campus, in terms of both enrollment and physical space (Inkelas and Weisman, 2003). They have been proposed to facilitate the transition to college through greater opportunities to interact with peers and faculty, as well as more applied modes of learning, such as internships, service learning, research projects, and so forth (Schoem, 2004). Finally, as stated previously, LLPs bridge the in-class and out-of-class experience for students, promoting a more seamless learning environment in addition to a more intimate one (Laufgraben, Shapiro, and Associates, 2004).

Living-Learning Programs Focusing Specifically on STEM Topics

It should be of little surprise, then, that LLPs became extremely popular in the late 1980s, through the 1990s, and into the twenty-first century, especially since pundits and policymakers continue to call for undergraduate education reform (see, for example, the Boyer Commission's *Reinventing Undergraduate Education: A Blueprint for America's Research Universities*, 1998).

There is currently no comprehensive listing of all LLPs in existence, but a 2007 data collection of the National Study of Living-Learning Programs (NSLLP) included more than fifty institutions and six hundred living-learning programs across the United States (Inkelas and Associates,

NEW DIRECTIONS FOR INSTITUTIONAL RESEARCH • DOI: 10.1002/ir

2007). However, those six-hundred-plus LLPs can be distilled into broad categories by theme, and the NSLLP researchers created a thematic typology of programs that span seventeen types: civic/social leadership, cultural, disciplinary, fine and creative arts, general academic, honors, leisure, political interest, research, Reserve Officer Training Corps (ROTC), residential colleges, sophomore, transition, umbrella, upper-division, wellness, and women's programs. In addition, several of the categories were further subdivided into subtypes, to form forty-one types in all. (For more information about the NSLLP thematic typology, see Inkelas and Soldner, 2011.)

According to the NSLLP findings, women in STEM majors could be found in nearly every type of LLP, from honors programs to transition programs to women's programs. However, there were several types of LLPs that catered specifically to women in STEM or all students with STEM interests. The first type is designed specifically for women in STEM majors, and only women participate in the program. The NSLLP identified fourteen such programs in its data collection, with the actual names of the programs including Women in Math, Science, and Engineering (WIMSE) at the University of Illinois and Florida State University (separately), or Women in Science and Engineering (WISE) at the University of Arizona, Clemson University, Syracuse University, the University of Wisconsin, New Mexico State University, and the University of Michigan (separately).

Second, there were three types of coeducational LLPs that focused on a specific aspect of STEM: (1) engineering and computer science, (2) health science, and (3) general science. Engineering and computer science LLPs tended to focus on those two disciplines, while general science programs included a variety of topics ranging from environmental sciences at the University of Colorado to human ecology at Ohio State University. The health sciences programs largely focused on medical fields, such as emergency health services at the University of Maryland, Baltimore County; nursing at the University of Missouri and Ohio State University; and sports and society at Indiana University. The NSLLP identified twenty-seven coeducational engineering and computer science LLPs, eighteen general science LLPs, and twenty health science LLPs. However, it is important to reiterate that women in STEM majors enrolled in all types of LLPs, including those with a non-STEM theme, such as honors programs, multicultural programs, and the like.

Empirical Evidence on Living-Learning Effectiveness and STEM Success for Women

What evidence is there that LLPs—regardless of the type of LLP—are effective in enhancing success for women in STEM? Unfortunately, there is not much empirical research on this topic, and of the limited number of studies available much of the information comes from the National Study of Living-Learning Programs. The NSLLP was administered to a sample of LLP and non-LLP students in spring 2007 at more than fifty colleges and universities across the United States. The LLP sample consisted of 11,606 respondents representing some six hundred living-learning programs. The non-LLP (or comparison) sample included 10,913 respondents who were matched to the LLP sample at each of the participating institutions by gender, race or ethnicity, academic class standing, and residence hall occupancy. A total of 22,519 LLP and non-LLP students completed the Residence Environment Survey online during the spring term in 2007; the precise dates for data collection varied by individual campus calendars' spring term start dates and spring breaks. The Residence Environment Survey included questions about students' background characteristics, their college experiences (including experiences in STEM majors as well as STEM or non-STEM-based LLPs), and their outcomes, or how they have been influenced by their college attendance (Inkelas and Associates, 2007). For information on the NSLLP and the Residence Environment Survey, see http://www.livelearnstudy.net.

The 2008 NSLLP report of findings from Inkelas and Associates revealed that women in STEM did experience different outcomes as a result of their participation in various types of LLPs. The report categorized women in STEM majors into four groupings: (1) those who participated in women-only STEM LLPs such as women in science and engineering programs; (2) those who participated in coeducational STEM LLPs such as engineering and computer science, health science, or general science programs; (3) those who participated in LLPs that did not focus specifically on STEM topics, such as honors, cultural, or fine/creative arts programs; and (4) those who did not participate in any form of LLP.

The findings below summarize empirical research on the relationship between LLP participation and various outcomes for women in STEM majors. Though the majority of the data reported are from the NSLLP, findings from other studies are also included.

Academic and Social Transition. Analyzing the NSLLP data, Inkelas and Associates (2007) found that women in STEM majors who were involved with women-only STEM LLPs or coeducational STEM LLPs were more likely to express a smooth social transition to college than women in STEM majors who did not participate in a LLP. However, only women in coeducational STEM LLPs expressed a smoother academic transition. Helman (2000) similarly found that women in a coeducational STEM-based LLP reported a successful academic adjustment to college, and a more successful adjustment than males in the LLP. Helman's study included 174 first-year men and women enrolled in a coeducational LLP called ROSES (Residential Option for Science and Engineering Students) who also declared their major in science or engineering in the fall of 1999 at Michigan State University.

NEW DIRECTIONS FOR INSTITUTIONAL RESEARCH • DOI: 10.1002/ir

Perceptions or Intellectual Abilities. Interestingly, Inkelas and Associates (2007) reported that women in non-STEM-based LLPs expressed using a greater level of critical thinking in their coursework than those in coeducational STEM LLPs or no program at all. However, the higher level of critical thinking scores among women in the non-STEM-based LLPs may be due to the fact that women in STEM majors in the NSLLP who are participants in those types of programs were predominantly clustered in honors programs, which tend to emphasize critical thinking in their programming. It seems likely that students in honors programs may exercise their critical-thinking skills at a higher level than students in other types of LLPs or no LLP at all.

Self-Confidence. Perceptions of confidence among women in STEM varied widely depending on LLP type, according to findings from the 2007 NSLLP (Inkelas and Associates, 2007). Women in non-STEM LLPs expressed the highest sense of confidence in their college success (again this may be an honors program effect) and test-taking skills; however, women in women-only STEM LLPs were the most likely to express confidence in their math or engineering courses. Meanwhile, behaviors of low self-confidence, such as dropping a class, doing less well in a particular class than expected, and feeling overwhelmed by homework, were most likely associated with women in STEM majors who did not participate in any form of LLP.

Academic Performance. The 2007 NSLLP revealed no significant differences in college grade point averages among women in STEM who did or did not participate in an LLP (Inkelas and Associates, 2007), but Helman (2000) found that women in a coeducational STEM LLP had significantly higher fall semester grade point averages than males in the same program. On the other hand, Allen (1999) reported that, for all three years of data she collected, women in a women-only STEM LLP (for that study, Women in Science and Engineering Residence Program, or WISE-RP) at the University of Wisconsin Madison had substantially higher first semester grade point averages than women in a control group or all first-year students at UW Madison. Data from the study included thirty-one WISE-RP students and sixty-five women in the control group, who enrolled in both mathematics and chemistry courses in their first year. The two groups were comparable in ACT score. Allen found the difference in grade point average for WISE-RP women in comparison to the all-first-year average at UW Madison all the more significant because of the high content of mathematics and science courses in the WISE-RP participants' courses taken, because these courses are generally associated with low grades.

Retention or Persistence. The 2007 NSLLP (Inkelas and Associates, 2007) showed no differences in drop-out risk among women in STEM in any of the three types of LLPs or the comparison group. However, Soldner, Rowan-Kenyon, and Inkelas (forthcoming) studied the role of LLP participation, among several factors, in supporting persistence in STEM majors for both men and women. Using the same NSLLP data from 2007, they found that participation in an LLP (with a STEM focus or otherwise) had no direct relationship with STEM persistence for women. However, women's likelihood of graduating from a STEM field was related to several facets of their college experiences that are generally assumed to be augmented through participation in LLPs, notably conversations with peers about academic issues, faculty interaction and mentoring, and socially supportive residence hall climates. Thus, even though mere participation in an LLP may not necessarily facilitate STEM persistence for women, the types of activities undertaken in LLPs (STEM-focused or otherwise) may be the conduit for STEM success for women.

An earlier study conducted on women participating in a women-only STEM LLP (the Women in Science and Engineering Residential Program, or WISE-RP) at the University of Michigan reported more nuanced effects on retention (Hathaway, Sharp, and Davis, 2001). There were 1,852 students in the Hathaway, Sharp, and Davis study's sample, 264 of whom were WISE-RP participants between 1993 and 1997. There were also two control groups, one consisting of women (n = 794) and one consisting of men (n = 794) who were science or engineering majors but not part of the WISE-RP. The three groups were also matched on high school grade point average, SAT or ACT score, and high school type. The authors found that WISE-RP women were retained in science majors at a significantly higher rate than their male and female counterparts in the control samples. However, there were no significant differences in retention rate in engineering among WISE-RP women, women in the control group, and men in the control group. Among the women changing their major from science originally to something else, WISE-RP students were more likely to change to a different science major than to leave the sciences altogether. Again, there were no similar patterns detected among engineering majors.

STEM Graduate School Plans. There is evidence that participating in LLPs is related to the likelihood of attending graduate school. A study using the NSLLP data yielded the finding that participating in a STEM-focused LLP was directly related to positive STEM-related outcomes for women. Szelenyi and Inkelas (2011) examined the longitudinal data from the NSLLP, which included 294 women STEM majors who responded to the baseline data collection in 2004 as first-year students and then again in the follow-up data collection in 2007, as fourth-year students. These women respondents were from sixteen postsecondary institutions across the United States, and nearly 59 percent of the respondents participated in some form of LLP during their first year of college in 2004. In their study, the authors found that women in STEM majors who participated in a women-only STEM LLP (e.g., the Women in Science and Engineering Program) were 35 percent more likely to attend graduate school in a STEM discipline than women in STEM majors who participated in a coeducational STEM LLP (e.g., Engineering House). Similarly, participants in women-only STEM LLPs were 31 percent more likely to attend graduate school in a STEM field than women STEM majors in non-STEM LLPs (e.g., honors programs), and 29 percent more likely than women in STEM who did not participate in any type of LLP. Thus single-gender LLP experiences appear to play a role in supporting women's graduate school plans in STEM. Even more compelling is the fact that, for most of the women in the study, participation in the women-only STEM LLP ended after their freshman year. That participation in any intervention in the first year is shown to boast a statistically significant influence on outcomes four years later is particularly notable, but this study also took into account the women's background characteristics and other college experiences, further strengthening confidence in the relationship between LLP participation and graduate school plans.

Other Outcomes. Allen (1999) also found that women in women-only STEM LLPs (for this study, women in WISE-RP at UW Madison) were more satisfied with their college experience and tended to binge-drink less than the women in the control group. Only 19 percent of WISE-RP women had consumed four or more alcoholic drinks in one sitting over a two-week period prior to the survey, while 72 percent of women in the control group reported drinking four or more alcoholic beverages during the same time period.

Summary of Empirical Findings. As discussed previously, there are very few studies that examine the roles and contributions of living-learning programs in facilitating successful outcomes for women in STEM majors. Other than works using NSLLP data, I found only three studies that examined this intervention and its effects on women in STEM. Thus what is currently understood about LLPs and women in STEM is only preliminary at best and must be replicated in future studies in order to uncover any patterns in findings. It is also important to note that the studies summarized here did not employ similar research designs or methods and consequently are difficult to compare. For example, even though the NSLLP examined women in STEM outcomes among participants in (1) women-only STEM LLPs, (2) coeducational STEM LLPs, (3) non-STEM LLPs, and (4) a control sample, other studies examined just women-only STEM LLPs (Allen, 1999; Hathaway, Sharp, and Davis, 2001) or coeducational STEM LLPs (Helman, 2000).

The limitations aside, there are some interesting findings about apparent differences in influence among the several types of LLPs. For example, women-only LLP participation has been associated with a successful social transition to college, confidence in mathematics and engineering courses, plans to attend graduate school, satisfaction with college, and lower binge drinking for women STEM majors. On the other hand, coeducational LLPs have been associated with successful academic transitions or adjustments. Finally, non-STEM-related LLPs are also associated with certain outcomes, including overall confidence in college success and use of critical-thinking skills.

Implications for Institutional Research

To continue studying the relationship between LLP participation and out-comes for women in STEM, here are several recommendations for institutional researchers to follow in paving the way for this type of inquiry. First, institutional record keeping should include a flag on students' records if they participated in any type of living-learning program. This should include every type of living-learning program, whether it be a women-only STEM LLP, a coeducational STEM LLP, or a non-STEM-based LLP, and also for both women and men. This type of inquiry is typically comparative, and thus to understand how women in STEM differ from their counterparts vis-à-vis living-learning programs, we must also know which women in STEM did not participate in an LLP, and which men did or did not participate.

Second, institutions should undergo ongoing tracking of academic performance and retention for women in STEM. As reviewed previously, there are currently empirical studies that contradict one another on whether women in STEM who participated in any form of LLP persist at an equal or higher rate or obtain higher grades than women in STEM (or men) who did not participate in an LLP. Additional evidence from diverse sets of institutions can help us establish what the prevailing patterns are regarding retention and academic performance.

Finally, researchers studying women in STEM success might broaden their conceptions of what encompasses "success" for such women in living-learning programs. Two of the more interesting findings from the literature were that women in STEM who participated in a women-only STEM LLP were significantly more likely to indicate pursuing graduate school in a STEM field, and they were significantly less likely to binge-drink. (It should be noted that a lower level of binge drinking is commonly found among living communities in which the residents felt stronger bonds with one another; Brower, 2008). These outcomes of LLP participation may not be central in the study of the STEM pipeline, but they do offer a glimpse into smaller success stories for women, who remain critically underrepresented at every step of the STEM journey.

Conclusion

Although living-learning programs were not founded and did not become popular in the past few decades because of their potential to improve outcomes for women in STEM disciplines, preliminary results from a few empirical studies suggest that LLPs may be effective in facilitating such success. Interestingly, both LLPs and women in STEM became part of American higher education's critical consciousness thanks to discontent from external stakeholders regarding the quality of undergraduate education and the pervasive underrepresentation of women in scientific fields.

How ironic it would be if the solution to both problems (i.e., poor quality in undergraduate education and underrepresentation of women in STEM) were to combine the two efforts.

References

Alexander, F. K. "From Oxbridge to the Colonies: The Development and Challenges of Residential Colleges in American Public Universities." In F. K. Alexander and D. E. Robertson (eds.), *Residential Colleges: Reforming American Higher Education*. Louisville, Ky.: Oxford International Round Table, 1998.

Allen, C. "WISER Women: Fostering Undergraduate Success in Science and Engineering with a Residential Academic Program." *Journal of Women and Minorities in Science and Engineering*, 1999, 5, 265–277.

Association of American Colleges and Universities. *Greater Expectations: A New Vision for Learning as a Nation Goes to College*. Washington D.C.: AAC&U, 2002.

Bloom, A. D. *The Closing of the American Mind: How Higher Education Has Failed Democracy and Impoverished the Souls of Today's Students*. New York: Simon & Schuster, 1987.

Boyer Commission on Educating Undergraduates in the Research University. *Reinventing Undergraduate Education: A Blueprint for America's Research Universities*. Stony Brook: State University of New York, 1998.

Brower, A. M. "More like a Home Than a Hotel: Living-Learning Programs Theme Issue." *Journal of College and University Student Housing*, 2008, 35(1), 32–49.

Chaddock, K. E. "From Inventions of Necessity to Necessary Invention: The Evolution of Learning in Residential Settings." In G. Luna and J. Gahagan (eds.), *Learning Initiatives in the Residential Setting*. The First-Year Experience Monograph Series, no. 48. Columbia: National Resource Center for the First-Year Experience and Students in Transition, University of South Carolina, 2008.

Gabelnick, F., MacGregor, J., Matthews, R. S., and Smith, B. L. (eds.). *Learning Communities: Creating Connections Among Students, Faculty, and Disciplines*. New Directions for Teaching and Learning, no. 41. San Francisco: Jossey-Bass, 1990.

Hathaway, R. S., Sharp, S., and Davis, C.-S. "Programmatic Efforts Affect Retention of Women in Science and Engineering." *Journal of Women and Minorities in Science and Engineering*, 2001, 7, 107–124.

Helman, C.K. "Adjustment for College: The Contribution of a Living-Learning Program for Science and Engineering Students." Doctoral dissertation, Michigan State University, 2000.

Inkelas, K. K., and Soldner, M. "Undergraduate Living-Learning Programs and Student Outcomes." *Higher Education: Handbook of Theory and Research*, 26, 2011.

Inkelas, K. K., and Weisman, J. L. "Different by Design: An Examination of Student Outcomes Among Participants in Three Types of Living-Learning Programs." *Journal of College Student Development*, 2003, 44(3), 335–368.

Inkelas, K. K., and Associates. *National Study of Living-Learning Programs: 2007 Report of Findings*. College Park, Md.: Inkelas and Associates, 2007. Retrieved from http://hdl.handle.net/1903/8392.

Laufgraben, J. L., Shapiro, N. S., and Associates. *Sustaining and Improving Learning Communities*. San Francisco: Jossey-Bass, 2004.

Meiklejohn, A. *The Experimental College*. Madison: University of Wisconsin Press, 2001. (Originally published 1932.)

Ravitch, D. and Finn, C. E. *What Do Our Seventeen Year Olds Know? A Report on the First National Assessment of History and Literature*. New York: HarperCollins, 1988.

Rudolph, F. *The American College and University: A History*. Athens: University of Georgia Press, 1990.

Ryan, M. B. "Residential Colleges: A Legacy of Living & Learning Together." *Change*, 1992, 24(5), 136–140.

Schoem, D. "Sustaining Living-Learning Programs." In J. L. Laufgraben and N. S. Shapiro (eds.), *Sustaining and Improving Learning Communities*. San Francisco: Jossey-Bass, 2004.

Shapiro, N. S., and Levine, J. H. *Creating Learning Communities: A Practical Guide to Winning Support, Organizing for Change, and Implementing Programs*. San Francisco: Jossey-Bass, 1999.

Soldner, M., Rowan-Kenyon, H., and Inkelas, K. K. "Supporting Persistence in STEM Disciplines: The Role of Living-Learning Programs Among Other Social-Cognitive factors." *Journal of Higher Education*, forthcoming.

Sykes, C. J. *ProfScam: Professors and the Demise of Higher Education*. Washington D.C.: Regnery Press, 1988.

Szelenyi, K., and Inkelas, K. K. "The Role of Living-Learning Programs in Women's Plans to Attend Graduate School in STEM Fields." *Research in Higher Education*, 2011, 52 (4), 349–369.

Veysey, L. R. *The Emergence of the American University*. Chicago: University of Chicago Press, 1965.

KAREN KUROTSUCHI INKELAS is an associate professor at the Curry School of Education at the University of Virginia, and director of the UVA Center for the Advanced Study of Teaching and Learning in Higher Education.

NEW DIRECTIONS FOR INSTITUTIONAL RESEARCH • DOI: 10.1002/ir

This chapter discusses the role of community colleges in educating the next generation of women in science, technology, engineering, and mathematics (STEM). Implications for policy and practice are offered, based on the major findings from a mixed-method study on the experiences of women in STEM who transferred to a four-year college from the community college system.

The Role of Community Colleges in Educating Women in Science and Engineering

Dimitra Lynette Jackson, Frankie Santos Laanan

The low representation of females in science, technology, engineering, and mathematics (STEM) fields remains a concern for the U.S. economy. Although America is still positioned with a disproportionate share of the world's finest universities—particularly research universities (National Academy of Sciences, 2010)—the reputation of the United States as having the most-prepared workforce in terms of STEM is being challenged because of the lack of individuals, specifically women, who are prepared to assume the workforce demand. According to the U.S. Department of Labor, by 2014 the fastest-growing occupations will require significant training in science and mathematics to successfully compete for a job (Jones, 2008). Consequently, the U.S. needs to ensure that individuals are prepared in the areas of math and science in order to meet the needs of the nation. Community colleges have been acknowledged as having the potential to play a key role in this increase of human capital (Starobin, Laanan, and Burger, 2010). Therefore, understanding the experiences of female students will promote educational environments that are conducive to student success in STEM fields.

Role of Community Colleges

As highlighted earlier in this volume, the current literature on gender differences in STEM reveals that a much smaller proportion of women than

New Directions for Institutional Research, no. 152, Winter 2011 © Wiley Periodicals, Inc.
Published online in Wiley Online Library (wileyonlinelibrary.com) • DOI: 10.1002/ir.407

men are pursuing careers in STEM. Community colleges have been identified as effective avenues for increasing the participation of women and underrepresented populations in STEM education (Berger and Malaney, 2003). Community colleges have many roles, among them providing academic preparation, vocational education, job training experiences, and programs catering to the community (Kasper, 2003). One essential role of the community college is the transfer function (Cohen and Brawer, 2008), where students transfer to a four-year institution to complete a bachelor's degree after attending a community college for the first two years of their education (Laanan, 1998). President Obama has recognized community colleges as being "uniquely positioned to raise the skill and knowledge base of our workforce . . . and in position to develop our nation's human capital" (White House, 2010b). Research supports the assertion that, compared to men, women scientists and engineers are more likely to have attended a community college at some point in their academic career (Tsapogas, 2004). Over the past few decades, community colleges have been recognized as key players in increasing the representation of female students in STEM areas (Starobin and Laanan, 2005).

Initiatives to Increase the Representation of Women in STEM

The Obama administration has put forth several initiatives designed to assist in increasing the strength of community colleges' and education's focus on STEM. Educate to Innovate (White House, 2010b) is one such initiative, designed to "improve the participation and performance of America's students in STEM." Through this campaign, the objectives are to "(1) increase STEM literacy, (2) move American students from the middle of the pack to the top in the next decade, and (3) expand STEM education and career opportunities for underrepresented groups, including women and girls." In addition, the Community College and Career Training Initiative (White House, 2010a) was established to award approximately $500 million to community colleges to assist in completing degrees and certificates. Moreover, through the American Recovery and Reinvestment Act (White House, 2010a) the Obama administration invested billions of dollars in community colleges, workforce training programs at community colleges, and work-study programs to assist students with the financial obligation of attending a community college.

In 2010, the White House hosted the inaugural White House Summit on Community Colleges. This was an opportunity to bring together community colleges and their students with business, philanthropy, federal, and state policy leaders to discuss how community colleges can help meet the job training and education needs of the nation's evolving workforce. A major topic of discussion was the role of community colleges in leading America to having the highest proportion of college graduates by 2020.

NEW DIRECTIONS FOR INSTITUTIONAL RESEARCH • DOI: 10.1002/ir

The summit highlighted foundations and educational organizations interested in investing in research and programs to assist in strengthening community colleges. The Aspen Prize for Community College Excellence announced a $1 million award to encourage and reward "outstanding outcomes in community colleges worldwide" for the purposes of sharing successful practices and highlighting "benchmarks for assessing community college outcomes" (White House 2010c). Hispanic serving institutions (HSI) are also gaining attention and federal support for STEM preparation efforts. HSI-STEM and Articulation Program (HSI-STEM) will receive an infusion of $100 million annually through 2019—a billion dollar investment—to increase degree attainment in STEM fields at HSIs (Malcolm, Dowd, and Yu, 2010).

In addition to federal initiatives, public and private partnerships have also been created to strengthen community colleges for the common goal of being the world leader in having the highest proportion of college graduates by 2020. One such partnership is the Gates Foundation: Completion by Design program. The aim of this program is to raise the stagnant completion rate of community college students by building on existing practices. This funding will assist in implementing model pathways to improve the graduation rate at community colleges (White House, 2010a).

Research on Women and Underrepresented Populations in STEM

Several studies have examined the role community colleges play in educating women and underrepresented populations in STEM fields. From this body of literature, a number of factors have been identified that have an impact on recruitment and retention of females in STEM fields. For example, positive and supportive academic environments and informal student interactions at the community college and university have been shown to be beneficial for community college students (Townsend and Wilson, 2006). Although academic preparation in math and science was found to positively affect the experiences of women in STEM (Oakes, 1990), structural and cultural academic factors such as chilly classrooms were also highlighted as factors that discouraged the participation of women and underrepresentation populations in STEM at four-year institutions (Hurtado, Carter, and Spuler, 1996). Mentee-mentor relationships and support systems (Creamer and Laughlin, 2005) at both types of institutions have also been identified as influential factors for students in STEM majors. A recent study conducted by Sax, Bryant, and Harper (2010) showed that student interactions with faculty were significant regarding self-confidence, retention, and degree aspirations. This finding is consistent with previous research (Astin, 1993; Kezar and Moriarty, 2000).

Laanan, Starobin, and Eggleston (2010) conducted a study that focused on the adjustment of community college transfer students. Their

NEW DIRECTIONS FOR INSTITUTIONAL RESEARCH • DOI: 10.1002/ir

study found that community colleges served as effective avenues for students to obtain transfer student capital (TSC) or effective learning and study skills. The TSC, defined by Laanan (Pappano, 2006; Laanan, Starobin, and Eggleston, 2010), is a conceptual lens to understand transfer students' socialization and development of knowledge, skills, and expertise. Specifically, TSC measures transfer students' ability to accumulate knowledge, information, and expertise to navigate the transfer process.

Additionally, studies conducted by Laanan and Starobin (2008) and Jackson (2010) support current research on female community college transfer students in STEM. In particular, Jackson (2010) found that women were interested in pursuing STEM areas and benefited from early encouragement and motivation to pursue STEM areas of study. Further, advising from faculty and program coordinators was critical in a student's decision to pursue a STEM area of study.

In 2010 Starobin and Laanan edited a special issue of the *Journal of Women and Minorities in Science and Engineering* highlighting the role of community colleges in broadening participation among women and minorities in STEM fields as well as their role in developing human capital in STEM areas. The issue addressed implications for policy, practice, and future research related to increasing the participation of women and underrepresented students in the STEM pathway.

Women currently account for at least half of the baccalaureate degrees awarded in most industrialized countries. They still, however, continue to be significantly underrepresented in science and technology (National Science Foundation, 1994; Stolte-Heiskanen, Acar, Ananieva, and Gaudart, 1991). Increasing the representation of females is essential in ensuring that there is a "diversity of perspectives in the search for knowledge and solutions to human problems" (Blickenstaff, 2005, p. 383).

Description of Study

The remainder of this chapter summarizes the findings from a study on the experiences of female STEM majors who transferred from a community college. The purpose of the study was to (1) understand the characteristics and experiences of female transfer students in STEM majors at a research institution; (2) present methodological strategies to better collect systematic data on community college women who pursue STEM disciplines and transfer to a four-year institution; and (3) inform faculty, administrators, and student affairs professionals in community college and university environments about strategies to recruit and retain female STEM students.

In our descriptive study, we used survey research techniques to investigate the background characteristics as well as the community college and university experiences of ninety-nine female community college transfer students in STEM majors at a Midwestern research university.

NEW DIRECTIONS FOR INSTITUTIONAL RESEARCH • DOI: 10.1002/ir

Additionally, this study sought to understand the factors that assisted in the adjustment of the female community college transfer students in STEM majors at the university.

Data were collected using the Laanan Transfer Student Questionnaire or L-TSQ (1998, 2004, 2007). The 132-item survey consists of questions about students' background characteristics, community college experiences, and university experiences. The survey also included questions regarding current place of residence, gender, age, ethnicity; highest academic degree intended to obtain, as well as parents' highest level of education and parents' total household income last year. The community college section of the survey asked students to respond to questions regarding their perception of general courses at the community college, and academic advising and counseling services at the community college. Lastly, the survey consisted of questions that highlighted their university experiences, which included items such as reasons for attending the university, as well as information regarding course learning, and experiences with faculty. One open-ended question asked the female students to discuss factors that assisted in their adjustment to a Midwestern research university.

Summary of Findings

Regarding participants' experiences at the community college, the majority (53.5 percent) did not complete an associate degree prior to transferring to the four-year institution. In terms of the transfer semester GPA, the largest proportion (39.4 percent) transferred with a GPA of 3.5 or higher. A little over one-third (35.4 percent) transferred with a GPA of 3.00–3.49. As for transfer semester hours, about three-fourths (75.8 percent) transferred with fifty to one hundred credit hours. Regarding time spent studying and preparing for class, more than three-fourths (78.1 percent) spent between one and ten hours studying per week while at the community college. In terms of academic advising and counseling services, more than three-fourths (76.4 percent) agreed somewhat or agreed strongly to having consulted or talked with advisors or counseling services about transferring.

Additionally, at the community colleges approximately three-fourths (75.6 percent) often or very often interacted with and felt comfortable approaching faculty. In relation to the course learning construct, a large majority (97.6 percent) often or very often took notes and engaged in class activities, projects, and class discussions. Regarding the general-courses construct, a large majority (88.0 percent) agreed somewhat or agreed strongly that general courses were intellectually challenging and extensive, and that they prepared them for university courses.

As related to the transfer process construct, the majority took a proactive approach to learning about the transfer institution. More than three-fourths (86.7 percent) agreed somewhat or agreed strongly to researching and visiting the university campus to learn more about the campus prior to transferring.

Descriptive statistics concerning participants' experiences at the four-year institution show that more than three-fourths (77.6 percent) lived on campus. Some one-third (38.8 percent) maintained a 2.00–2.99 GPA at the university as of fall 2009. About two-thirds (66.0 percent) decided to attend the university to obtain a bachelor's degree. One-fifth (20.5 percent) attended the university to pursue graduate or professional school. Slightly less than half (49.4 percent) found community college faculty, friends, and the university academic advisors to be somewhat important in their decision to attend the university. Concerning financial reasons for attending the university, 80.0 percent found financial issues to be important or very important. More than three-fourths (77.0 percent) found the reputation of the university construct to be important or very important. In terms of the university-sponsored transfer student orientation, 88.9 percent participated. The majority (70.4 percent) found the transfer student orientation to be somewhat helpful or very helpful. In relation to the course learning construct, almost all respondents (93.8 percent) often or very often took notes and participated in class discussions and activities.

The participants in the study reported good experiences with faculty at their four-year institution. Over half (58.8 percent) often or very often visited faculty or felt comfortable interacting with faculty. In terms of the general perceptions of faculty construct, 90.2 percent agreed somewhat or agreed strongly to faculty being easy to approach, responsive, and concerned with their academic progress.

As for the perception of transfer students, 67.5 percent agreed somewhat or agreed strongly that the abilities of transfer students were underestimated and that there was a stigma or negative perception of transfer students. In terms of overall satisfaction, 90.2 percent agreed somewhat or agreed strongly that the courses at the university were interesting and the campus was intellectually stimulating.

Summary of Qualitative Findings

The open-ended question on the survey allowed female community college transfer students to comment on factors that assisted in their adjustment to the university. Three major themes emerged: (1) transfer student orientation, (2) assistance from advisors, and (3) getting involved on campus.

Transfer student orientation. The majority of students highlighted the benefit of the student transfer orientation in their adjustment to the university. The students also reported how the transfer orientation provided avenues for meeting other transfer students:

"The orientation program for transfer students was really helpful."

"I think my orientation classes with sections specifically for transfer students were very helpful. I liked that I met other transfer students who were a little closer to my age and in my major. I've had several of these students in subsequent classes."

Advisors. In addition to the transfer student orientation, the students valued interaction with their advisor and found the interaction to be an essential component of the transfer process. The female transfer students reported how their advisors were able to assist them with class selection, answer questions, and help them feel comfortable with the transfer process. The students also drew attention to how assistance from their advisors was the very reason they decided to enroll at the university:

"When I asked some questions about transfer to a Midwestern research university advisor, she answered my questions kindly, and she could make the time to talk with me. Her help became one of the factors which I decided to go to a Midwestern research university."

"My advisor made my transfer easy, pain-free, and quick. Honestly, if it weren't for her, I might not have enrolled when I did."

Getting involved. The female community college transfer students valued getting involved in campus organizations, clubs, and welcome events. Some of the clubs and organizations in which the students were involved were the Farm Op Club, Student Veterans of America Club, Destination Team Leader, sororities, and clubs and organizations within the individual's major. The students viewed this affiliation as a way to feel connected to the college environment. Being involved on campus was viewed as a great way to assist in the adjustment to a Midwestern research university:

"Getting involved definitely helped me become adjusted to a Midwestern research university. The different clubs helped me meet new people."

"I joined a sorority which has helped immensely when it comes to making friends and meeting new people. Also, my on-campus job helped to further me in my major and to make friends."

In addition to the clubs and organizations, one student expressed the need for a push for students to get involved on campus. The students highlighted how, although involvement is an important factor in the transfer process, it can be difficult to take the necessary steps to get involved when a student does not feel accepted or connected to the environment:

NEW DIRECTIONS FOR INSTITUTIONAL RESEARCH • DOI: 10.1002/ir

"It would have been great if there was more push for involvement around campus. It's hard to take the step to get involved in a group and even harder when you don't feel like you're accepted."

Conclusions

The findings from this study indicate that female students are entering educational systems with aspirations of obtaining at least a bachelor's degree in a STEM discipline. A fundamental question still remains: How can community college and four-year institutions be made conducive for female students to begin and successfully complete a STEM bachelor's degree? The female students in the study viewed the community college environment as an intellectual and comfortable environment. At the community college, the majority of the female students felt comfortable talking with faculty, were engaged in classroom discussions and activities, and viewed the courses as challenging. Although most students did not complete an associate degree prior to transferring, the majority of the ninety-nine female transfer students in STEM transferred with at least a 3.0 GPA and transferred with a minimum of fifty to one hundred credit hours. Additionally, the women in the study were proactive in researching prospective universities prior to transferring and consulted with an academic advisor before transferring to a university to discuss transfer options and required transfer courses. In addition to their academic progress, the female students found it beneficial to be engaged in extracurricular activities on campus.

The findings also indicated that the university environment had an impact on the experiences and aspirations of the female students in STEM areas. The majority of the female transfer students continued to actively participate in class discussions and activities and took class notes at the university. However, although this study found that students often or very often felt comfortable talking to faculty at the community college, the university experiences with faculty were a little different. The female students at the university felt that university faculty were interested in their learning and were concerned about their academic progress, but students did not feel as comfortable talking to and approaching faculty as they did at the community college.

Interestingly, the majority of the female transfer students agreed somewhat to feeling that their abilities were underestimated and that a stigma and negative perceptions of transfer students existed at the university. However, the female students viewed the university as very interesting and intellectually stimulating.

It is important to note that the findings in this study represent the experiences of traditional-age, white students. This study does not represent the experiences of students of color or nontraditional-age students, which may vary. The final section of the chapter offers recommendations for institutional research and directions for future study on this topic.

Implications for Institutional Research

Several implications for institutional research can be derived from the findings of our study. Examining the extent to which the experiences of women in STEM fields who transferred from a community college are similar to nontransfer women and nontransfer men in STEM would shed light on the factors that directly influence the experiences of transfer students. This knowledge would assist in understanding varying experiences regarding the transfer component. In addition to conducting quantitative studies, it would also be beneficial to conduct focus groups and interviews with those who leave to gain a deeper understanding of the challenges and factors having an impact on their decision to discontinue pursuing a degree in STEM. Further investigating transfer students' perceptions of the stigmas that exist regarding transfer students would shed light on additional factors that affect the experiences of community college transfer students. These findings would assist in understanding ways to have a positive impact on the experiences of community college transfer students in STEM majors.

As mentioned earlier, the L-TSQ is a survey instrument consisting of questions regarding the students' background experiences as well as their community college and university experiences. Understanding the experiences of female community college transfer students assists in understanding how students are making meaning of the environments to which they are being exposed. Merging the L-TSQ with academic transcript-level data (for example, number of credits in STEM, number of credits, grades in course, etc.) will assist in understanding the holistic experiences of female community college transfer students in STEM. Understanding the academic preparation component and the number of credits and grades obtained in math and science courses would add to the literature about student academic preparation.

Continuing to assess the classroom and campus dynamics would also be beneficial in understanding how the community college and university classroom environments influence student success in STEM majors. Future research that intentionally examines the experiences of nontraditional female community college transfer students in STEM majors would assist in understanding how their experiences are both similar to and different from those of their traditional counterparts. This will assist in understanding how to work with all community college transfer students. Particular attention should focus on the experiences of students of color and their experiences in STEM majors as well.

Community colleges have been highlighted as not only a pathway for students of varying backgrounds but also an entity positioned to develop the skill set and knowledge level of our current and future workforce. The experiences of these students both at the community college and the university as well as their adjustment to the university are essential in ensuring that students are able to successfully complete their degree in a STEM major. The success of all community college transfer students, particularly in STEM areas, means success for America's economy.

References

Astin, A. W. *What Matters in College? Four Critical Years Revisited.* San Francisco: Jossey-Bass, 1993.

Berger, J. B., and Malaney, G. D. "Assessing the Transition of Transfer Students from Community Colleges to a University." *NASPA Journal,* 2003, *40*(4), 1–23.

Blickenstaff, J. C. "Women and Science Careers: Leaky Pipeline or Gender Filter?" *Gender and Education,* 2005, *17*(4), 369–386.

Cohen, A. M., and Brawer, F. B. *The American Community College* (5th ed.). San Francisco: Jossey-Bass, 2008.

Creamer, E. G., and Laughlin, A. "Self-Authorship and Women's Career Decision-Making." *Journal of College Student Development,* 2005, *46*, 13–27.

Hurtado, S., Carter, D. F., and Spuler, A. "Latino Student Transition to College." *Research in Higher Education,* 1996, *37*(2), 135–157.

Jackson, D. L. "Transfer Students in STEM Majors: Gender Differences in the Socialization Factors That Influence Academic and Social Adjustment." Unpublished doctoral dissertation, Educational Leadership and Policy Studies, Iowa State University, 2010.

Jones, R. B. "Science, Technology, Engineering and Math SETDA." 2008. Retrieved August 1, 2009, from http://www.setda.org/c/document_library/get_file?folderId=270&name=DLFE-257.pdf.

Kasper, H. T. "The Changing Role of Community Colleges." *Occupational Outlook Quarterly,* 2003, *46*, 14–21.

Kezar, A., and Moriarty, D. "Expanding Our Understanding of Student Leadership Development: A Study Exploring Gender and Ethnic Identity." *Journal of College Student Development,* 2000, *41*, 55–69.

Laanan, F. S. "Beyond Transfer Shock: A Study of Students' College Experiences and Adjustment Processes at UCLA." Unpublished doctoral dissertation, Graduate School of Education and Information Studies, University of California, Los Angeles, 1998.

Laanan, F. S. "Studying Transfer Students: Part I: Instrument Design and Implications." *Community College Journal of Research and Practice,* 2004, *28*, 331–351.

Laanan, F. S. "Studying Transfer Students: Part II: Dimensions of Transfer Student Adjustment." *Community College Journal of Research and Practice,* 2007, *31*, 37–59.

Laanan, F. S., and Starobin, S. "Broadening Female Participation in Science, Technology, Engineering, and Mathematics: Experiences at Community Colleges." In J. Lester (ed.), *Gendered Perspectives on Community Colleges.* New Directions for Community Colleges, no. 142. San Francisco: Jossey-Bass, 2008.

Laanan, F. S., Starobin, S., and Eggleston, L. E. "Adjustment of Community College Students at a Four-Year University: Role and Relevance of Transfer Student Capital for Student Retention." *Journal of College Student Retention,* 2010, *12*(2), 175–209.

Malcolm, L. E., Dowd, A. C., and Yu, T. "Tapping HSI-STEM Funds to Improve Latina and Latino Access to the STEM Professions." Los Angeles: University of Southern California, 2010.

National Academy of Sciences. "Rising Above the Gathering Storm, Revisited: Rapidly Approaching Category 5." 2010. Retrieved August 2, 2010, from http://www.nap.edu/catalog.php?record_id = 12999.

National Science Foundation. *Women and Minorities in Science and Engineering: An Update.* Arlington, Va.: National Science Foundation, 1994.

Oakes, J. "Opportunities, Achievement, and Choice: Women and Minority Students in Science and Mathematics." *Review of Research in Education*, 1990, *16*, 153–222.

Pappano, L. "College, My Way: Lost, Alone, and Not a Freshman." *The New York Times*, April 23, 2006.

Sax, L. J., Bryant, A. N., and Harper, C. E. "The Differential Effects of Student-Faculty Interaction on College Outcomes for Women and Men." *Journal of Student Development*, 2010, *46*(6), 642–657.

Starobin, S. S., and Laanan, F. S. "Influence of Pre-College Experience on Self-Concept Among Community College Students in Science and Engineering." *Journal of Women and Minorities in Science and Engineering*, 2005, *11*(3), 209–230.

Starobin, S. S., Laanan, F. S., and Burger, C. L. "From Community College to Ph.D.: Educational Pathways in Science, Technology, Engineering, and Mathematics. *Journal of Women and Minorities in Science and Engineering*, 2010, *16*(1), 1–5.

Stolte-Heiskanen, V., Acar, F., Ananieva, N., and Gaudart, D. *Women in Science: Token Women or Gender Equality.* Oxford: Berg, 1991.

Townsend, B. K., and Wilson, K. B. "'A Hand Hold for a Little Bit': Factors Facilitating the Success of Community College Transfer Students to a Large Research University." *Journal of College Student Development*, 2006, *47*(4), 439–456.

Tsapogas, J. *The Role of Community Colleges in the Education of Recent Science and Engineering Graduates.* Washington, D.C.: National Science Foundation, 2004.

White House. "Building American Skills by Strengthening Community Colleges." 2010a. Retrieved November 20, 2010, from http://www.whitehouse.gov/sites/default/files/uploads/White-House-Summit-on-Community-Colleges-Fact-Sheet-100510.pdf.

White House. "Educate to Innovate." 2010b. Retrieved November 21, 2010, from http://www.whitehouse.gov/issues/education/educate-innovate.

White House. "White House Summit on Community Colleges." 2010c. Retrieved November 20, 2010, from http://www.whitehouse.gov/communitycollege.

DIMITRA LYNETTE JACKSON *is an Assistant Professor in the Department of Educational Psychology and Leadership at Texas Tech University. Her research focuses on the role and impact of community colleges in student success. She is also interested in student success and achievement in Science, Technology, Engineering and Mathematics (STEM) areas.*

FRANKIE SANTOS LAANAN *is an associate professor in the Department of Educational Leadership and Policy Studies at Iowa State University. His research focuses on the impact of community colleges on individuals and society.*

Current research on undergraduate women continues to provide insight on what dissuades, attracts, and retains them in STEM programs; however, less is known about their postbaccalaureate goals. Using social cognitive career theory (SCCT) as the theoretical framework, this chapter discusses the relationships among self-efficacy, the college environment, and women's postbaccalaureate goals.

The Postbaccalaureate Goals of College Women in STEM

Darnell Cole, Araceli Espinoza

For more than three decades, women have outnumbered men in college enrollments, yet they are less likely to graduate and enter a career in science, technology, engineering, or mathematics (STEM; National Center for Education Statistics, 2011). Even though women have made notable gains in social sciences, psychology, and biological and agricultural sciences (Babco and Ellis, 2006), significant disparities still persist in certain science and engineering fields. A study sponsored by the American Association of University Women (2010) titled *Why So Few?* indicates that in "physics, engineering, and computer science, the difference is dramatic, with women earning only 20 percent of bachelor's degrees" (p. xiv). The proportion of women who earn a doctorate in physics, engineering, or computer science is also approximately 20 percent. In terms of careers, women account for only 13.9 percent of physicists and astronomers, 10.6 percent of engineers, and 30 percent of computer scientists (American Association of University Women, 2010). The low number of women with a STEM bachelor's degree and with a STEM career is, according to some scholars, a product of the "leaky pipeline," a metaphor describing the educational conduit "carrying students from secondary school through [the] university and on to a job in STEM" (Blickenstaff, 2005, p. 369).

Theoretical Explanations

Research that examines the "leaky pipeline" for women in STEM fields often attributes the lack of women's entry into these fields to attitudes and early experiences (Blickenstaff, 2005), gender stereotypes and bias (Brainard and Carlin, 1998), academic preparation (Cole, 1997), a narrow overview of the field in introductory classes (American Association of University Women, 2010; Seymour and Hewitt, 1997), curriculum and pedagogy (Blickenstaff, 2005), lack of mentoring (American Association of University Women, 2010), and a chilly climate in science classrooms (Blickenstaff, 2005). Of these reasons used to explain the paucity of women in STEM fields, perhaps the most popular is the notion of the "chilly climate." For example, the *Why So Few?* report (American Association of University Women, 2010) specifies that the educational environment is in large part to blame for women's underachievement and disinterest in STEM careers. Parallel to the findings of the report, there are studies indicating that the academic context is likely to exert an influence on women's academic success and willingness to persist. In fact, women in STEM fields are more likely to be academically successful if they (1) establish rapport with faculty (Johnson, 2007), (2) have faculty respect (Huang and Brainard, 2001; Sax, Bryant, and Harper, 2005), (3) have faculty who care about them as individuals (Johnson, 2007), (4) have peers who believe they are capable in practical matters such as laboratory work (Seymour and Hewitt, 1997), and (5) enjoy adequate support when confronted with challenging classroom dynamics (Vogt, Hocevar, and Hagedorn, 2007).

Conversely, scholars who employ a self-efficacy perspective suggest that students' self-efficacy is perhaps more important than the educational environment and any individual barriers alone (Zeldin, Britner, and Pajares, 2008). From such a perspective, a woman's choice of a STEM major and her academic performance are anchored in her perceptions about her academic and intellectual abilities. For instance, there is evidence to indicate that even with good high school grades, if a woman lacks confidence in her academic ability she is less likely to report success in her STEM major (Huang and Brainard, 2001; Sax, 1994; Sax, Bryant, and Harper, 2005). Zeldin and Pajares (2000) also report that self-efficacy beliefs are important determinants of women's goals and their success in STEM careers.

Still other scholars, such as Vogt, Hocevar, and Hagedorn (2007), who rely heavily on Bandura's triadic reciprocal model (1986) of causality of the self (self-beliefs), the environment (interactions and perception of faculty and peers), and the behavior (persistence in STEM), find that women's pursuit of a STEM degree is significantly influenced by their perceptions of both the environment and their ability to succeed within that environment.

Social Cognitive Career Theory as an Explanation

Social cognitive career theory emphasizes the interplay of three social cognitive variables through which individuals regulate their own career development: self-efficacy, outcome expectations, and personal goals (Lent and Brown, 1996; Lent, Brown, and Hackett, 1994; Lent, Hackett, and Brown,

1996). Similar to Bandura's (1986) triadic reciprocal model of causality, the variables of social cognitive career theory—self-efficacy, outcome expectations, and goals—operate in conjunction with the person's context as well as individual quai..es such as sex or ethnicity, and socioeconomic status (Lent and Brown, 1996; Lent, Brown, and Hackett, 1994; Lent, Hackett, and Brown, 1996).

Self-efficacy refers to an individual's belief about his or her capabilities as opposed to what is objectively true (Bandura, 1986). For example, even if students are capable of completing the task at hand, if they question their own ability (low self-efficacy) they will perceive themselves as incapable (Schunk and Pajares, 2005). Outcome expectations involve the imagined product of acting in a particular manner (for example, "If I study, I will graduate with a good grade point average and I will have a high paying job"; Lent and Brown, 1996; Lent, Brown, and Hackett, 1994; Lent, Hackett, and Brown, 1996). Outcome expectations that may affect career behavior are anticipation of physical (monetary), social (approval), and self-evaluative (self-satisfaction) outcomes (Bandura, 1986). A goal is defined as the determination to engage in a particular activity to affect a particular future outcome (Bandura, 1986). Concepts such as career plans, decisions, aspirations, and expressed choices are all goal mechanisms (Lent and Brown, 1996; Lent, Brown, and Hackett, 1994; Lent, Hackett, and Brown, 1996).

For the study presented in this chapter, the occupational choice model of social cognitive career theory is used to interpret the postsecondary goals of undergraduate college women with STEM majors. In the occupation choice model, behavior is influenced by self-efficacy and outcome expectations, while choice is guided by whether individuals think they are competent (self-efficacious) and whether the expected outcomes (for example, salary) are worth the attempt (Lent and Brown, 1996; Lent, Brown, and Hackett, 1994; Lent, Hackett, and Brown, 1996). Whether individuals succeed or fail validates or nullifies their sense of self-efficacy or outcome expectations. Choice is therefore a recycling process that depends on how individuals perceive their capabilities. Choice is also constrained by contextual factors. For example, a woman's STEM career interest is more likely to develop into a goal if she perceives few barriers and a supportive environment. Finally, choice—as it is related to self-efficacy, outcome expectations, and goals—is influenced by such factors as socioeconomic status, educational access, social supports, gender role socialization, and community and family norms (Lent and Brown, 1996; Lent, Brown, and Hackett, 1994; Lent, Hackett, and Brown, 1996).

Description of Study

In this chapter we describe the results from a study that employed social cognitive career theory to examine how social cognitive and

NEW DIRECTIONS FOR INSTITUTIONAL RESEARCH • DOI: 10.1002/ir

environmental factors contribute to women's goals of either pursuing a STEM career or attending graduate school. Two research questions guided the study: What social cognitive and college environmental factors have an impact on the goal of pursuing a STEM career? What social cognitive and college environmental factors affect the goal of attending graduate school in any field?

With data from the Cooperative Institutional Research Program (conducted through the Higher Education Research Institute), we used binary logistic regressions to examine the influence of social cognitive and environmental factors on the postbaccalaureate goals for a sample of 1,804 undergraduate female students. In line with social cognitive theory, to account for self-efficacy we used a composite factor as a proxy, which included six items measuring math, writing, and academic ability, as well as drive to achieve and intellectual and social self-confidence. For outcome expectations we identified the importance of being well off financially, making a theoretical contribution, gaining recognition from colleagues, and helping others in need. Given that social cognitive career theory considers the context of the individual, there were seven college environmental factors included in this study. Of the seven environmental predictors, one was a composite to account for academic support from faculty. This composite included five items: opportunity to work on a research project, opportunity to publish, assistance with study skills, advice about educational program, and opportunity to discuss coursework outside of class. The remaining contextual predictors were encouragement for graduate school from faculty, respect from faculty, emotional support and encouragement from faculty, help from faculty in achieving professional goals, satisfaction with peer interactions, and satisfaction with courses in major field. Finally, the dependent variables—a woman's intent to pursue a STEM career or attend graduate school after graduation—were the goals. In sum, the identified variables for self-efficacy, outcome expectations, and the college environment were hypothesized to shape a woman's behavior in setting the goal of a STEM career or attending graduate school.

Summary of Findings

The regression results for the model that examined the influence of social cognitive and environmental factors on the goal of a "STEM career" revealed that the model was statistically reliable. Being a racial ethnic minority female, having good average high school grades, and parents having a higher income were all positively associated with reporting the goal of a STEM career. In addition, high self-efficacy beliefs, valuing making a theoretical contribution, and gaining recognition from colleagues were all positively associated with reporting the goal of a STEM career. In regard to the environmental predictors, being satisfied with the courses in their major field, reporting frequent faculty academic support, help from faculty in achieving professional goals, and respect from faculty were all positively associated with the likelihood of reporting the goal of a STEM career. Conversely, women who reported frequent emotional support and encouragement from faculty were less likely to report the goal of a STEM career.

The regression results for the model that examined the goal of "graduate school attendance in any field" indicated that the model was also statistically reliable, but not as many factors were significant in this model compared to the model that predicted the goal of a STEM career. However, higher parental income, having high self-efficacy beliefs, gaining recognition from colleagues, and reporting frequent encouragement for graduate school from faculty were all positively associated with the goal of attending graduate school. Finally, similar to the results for the goal of "STEM career," women who reported frequent emotional support and encouragement from faculty were less likely to report the goal of attending graduate school.

Conclusions and Recommendations for Institutional Research

Three major conclusions can be drawn from the results of this study that have implications for institutional research. First, the model for the goal of STEM career and the model for the goal of graduate school attendance were statistically reliable. There were, however, several more statistically significant predictors for the regression for the goal to pursue a STEM career than the regression for the goal to attend graduate school. On the basis of the predictors included in the logistic regression models, women's STEM career goals appear more malleable, which suggests that there are more predictors affecting women's goal of a STEM career than graduate school attendance. As a result, there are more opportunities that can be leveraged by faculty and other institutional agents toward increasing women's goals of a STEM career. The implication for institutional researchers, however, is that further research is needed to determine the extent to which such conclusions can be generally supported.

Second, there were three predictors significant toward women's goals of pursuing a STEM career or attending graduate school—self-efficacy, gaining recognition from colleagues, and emotional support and encouragement from faculty—although frequent emotional support and encouragement from faculty was negatively associated with the outcomes explored in this study. For instance, as a student's self-efficacy increased, so did the likelihood of her goal to pursue a STEM career or attend graduate school. Even after considering average high school grades, the findings also indicated that self-efficacy was significant toward a goal of STEM career or graduate school attendance. These findings are consistent with prior research, where (1) self-efficacy beliefs are important determinants of women's goals, choices, and success in STEM careers (Zeldin and Pajares, 2000; Sax, 1994; Sax, Bryant, and Harper, 2005); and (2) self-efficacy is meaningful in mediating academic performance (average high school grades) and future behaviors (the goal of a STEM career or graduate school attendance; Vogt, Hocevar, and Hagedorn, 2007; Zeldin, Britner, and Pajares, 2008).

It was not surprising that the desire to gain recognition from colleagues was significant for women's STEM careers or graduate school attendance, considering the increased success of women in STEM when they have peers who believe they are capable of being successful (Seymour and Hewitt, 1997). Research from Zeldin and Pajares (2000) regarding the influence of persuasions (social and verbal) in the construction of self-efficacy beliefs supports the notion that recognition from others is important to women's STEM-related goals, as well as attending graduate school. On the contrary, women who received frequent emotional support and encouragement from faculty were less likely to pursue either goal. This finding is in stark contrast to much of the literature on women in STEM, which suggests that faculty support and having faculty who care about them as individuals increases their academic performance (Johnson, 2007). In fact, establishing rapport with faculty (Johnson, 2007), having adequate support when confronted with challenging classroom dynamics (Vogt, Hocevar, and Hagedorn, 2007), and earning faculty respect (Huang and Brainard, 2001; Sax, Bryant, and Harper, 2005) have been positively correlated with women's success in STEM. Although earning faculty respect has also been identified in the current study as positively affecting a goal of a STEM career, little research exists to explain why emotional support and encouragement negatively affect a woman's goal of a STEM career or graduate school attendance. As a result, institutional researchers are encouraged to explore the relationships between the emotional support and encouragement received from faculty and a woman's goal of a non-STEM career.

Finally, social cognitive career theory (SCCT) appears to be a useful theoretical model for examining women's goals of a STEM career or attending graduate school. That is, women's behaviors are influenced by self-efficacy, outcome expectations, and contextual factors. Similarly, there is interplay among self-efficacy, outcome expectations, and contextual factors in women's goals of a STEM career or graduate school. As theorized earlier in this study, a woman's occupational choice is guided by whether she believes she is competent (self-efficacy) and whether the expected outcomes (recognition from colleagues) are worth the attempt of pursuing STEM-related goals (Lent and Brown, 1996; Lent, Brown, and Hackett, 1994; Lent, Hackett, and Brown, 1996). Within the context of the current study, a woman's goal of a STEM career or graduate school attendance is constrained only by emotional support and encouragement from faculty.

In sum, we hypothesized from prior literature that social cognitive and college environmental factors would significantly predict women's goals to pursue a STEM career or attend graduate school. The findings from our study indicate that social cognitive career theory is a useful theoretical and empirical model for predicting women's goals to pursue a STEM career or attend graduate school. Women who report high self-efficacy, a desire to gain recognition from colleagues, and low level of emotional support from faculty are more likely to report the goals of a STEM career or graduate school attendance. The implications for institutional researchers are to collect data and conduct analyses that consider women's self-efficacy, outcome expectations, and the contextual factors that are associated—in particular, their satisfaction within STEM major fields of study and their interactions with faculty—as important constructs for women's self-efficacy, outcome expectations, and STEM-related goals.

References

American Association of University Women. "Why So Few? Women in Science, Technology, Engineering, and Mathematics, 2010." Retrieved March 12, 2011, from http://www.aauw.org/learn/research/upload/whysofew.pdf.

Babco, E., and Ellis, R. Science and Technology Salaries: Trends and Details, 1995–2005. Commission on Professionals in Science and Technology, 2006. Retreived Oct. 21, 2011, from http://www.cpst.org/web/site/pages/pubs/PubOnline.cfm.

Bandura, A. Social Foundations of Thought and Action: A Social Cognitive Theory. Upper Saddle River, N.J.: Prentice-Hall, 1986.

Blickenstaff, J. C. "Women and Science Careers: Leaky Pipeline or Gender Filter?" Gender and Education, 2005, 17(4), 369–386.

Brainard, S. G., and Carlin, I. "A Six-Year Longitudinal Study of Undergraduate Women in Engineering and Science." Journal of Engineering Education, 1998, 87(4), 17–27.

Cole, N. S. The ETS Gender Study: How Females and Males Perform in Educational Settings. Princeton, N.J.: Educational Testing Service, 1997.

Huang, P., and Brainard, S. "Identifying Determinants of Academic Self-Confidence Among Science, Math, Engineering, and Technology Students." Journal of Women and Minorities in Science and Engineering, 2001, 7(4), 315–337.

Johnson, A. C. "Unintended Consequences: How Science Professors Discourage Women of Color." Science Education, 2007, 91(5), 805–821.

Lent, R. W., and Brown, S. D. "Social Cognitive Approach to Career Development: An Overview." Career Development Quarterly, 1996, 44(4), 310–321.

Lent, R. W., Brown, S. D., and Hackett, G. "Toward a Unifying Social Cognitive Theory of Career and Academic Interest, Choice, and Performance." (Monograph). Journal of Vocational Behavior, 1994, 45, 79–122.

Lent, R. W., Hackett, G., and Brown, S. D. "A Social Cognitive Framework for Studying Career Choice and Transition to Work." Journal of Vocational Education Research, 1996, 21(4), 3–31.

National Center for Education Statistics. "Digest of Education Statistics, 2011." Retrieved April 4, 2011, from http://nces.ed.gov/pubs2011/2011015.pdf.

Sax, L. J. "Retaining Tomorrow's Scientists: Exploring the Factors That Keep Male and Female College Students Interested in Science Careers." Journal of Women and Minorities in Science and Engineering, 1994, 1, 45–61.

Sax, L. J., Bryant, A. N., and Harper, C. E. "The Differential Effects of Student-Faculty Interaction on College Outcomes for Women and Men." *Journal of College Student Development*, 2005, *46*(6), 642–659.

Schunk, D. H., and Pajares, F. "Competence Perceptions and Academic Functioning." In A. J. Elliot and C. S. Dweck (eds.), *Handbook of Competence and Motivation.* New York: Guilford, 2005.

Seymour, E., and Hewitt, N. M. *Talking About Leaving: Why Undergraduates Leave the Sciences.* Boulder, Colo.: Westview Press, 1997.

Vogt, C. M., Hocevar, D., and Hagedorn, L. S. "A Social Cognitive Construct Validation: Determining Women and Men's Success in Engineering Programs." *Journal of Higher Education*, 2007, *78*(3), 336–364.

Zeldin, A. L., Britner, S. L., and Pajares, F. "A Comparative Study of the Self-Efficacy Beliefs of Successful Men and Women in Mathematics, Science, and Technology Careers." *Journal of Research on Science Teaching*, 2008, *45*(9), 1036–1058.

Zeldin, A. L., and Pajares, F. "Against the Odds: Self-Efficacy Beliefs of Women in Mathematical, Scientific, and Technological Careers." *American Educational Research Journal*, 2000, *37*, 215–246.

DARNELL COLE is an associate professor in the Rossier School of Education at the University of Southern California (USC) and also serves as the Ph.D. program chair for the Rossier School of Education.

ARACELI ESPINOZA is a Ph.D. student in the Rossier School of Education at USC.

This chapter examines the retention and degree completion of doctoral women in the science fields. The research shows that the lower degree completion rate for women may be a result of fewer opportunities for research assistantships.

Understanding the Factors Affecting Degree Completion of Doctoral Women in the Science and Engineering Fields

Frim D. Ampaw, Audrey J. Jaeger

The rate of doctoral degree completion, compared to all other degrees, is the lowest in the academy, with only 57 percent of doctoral students completing their degree within a ten-year period (Bell, 2008). In the science, engineering, and mathematics (SEM) fields, 62 percent of the male students complete their doctoral degree in ten years, which is better than the national average. For their female counterparts, however, the ten-year degree completion rate is 54 percent. Females in humanities and social sciences have a higher completion rate than males (Bell, 2008); yet they are lagging behind in the science, engineering and mathematics fields.

Women at the doctoral level are increasing their presence in the sciences but are not yet on par with their male counterparts. Moreover, the number of doctoral degrees awarded to females in the science and engineering fields has risen. In 2008, 41 percent of doctoral degree recipients in those fields were women compared to 36.7 percent in 2000 and 30 percent in 1994 (National Science Foundation, 1995, 2010). However, these higher numbers are driven by female doctoral degree recipients in psychology and other social sciences fields, as these physical sciences, life sciences, and social sciences are all in one group, where women represent 70 percent and 49 percent respectively of the doctoral student population (National Science Foundation, 2010). In regard to enrollment, women represent 36 percent of students entering science and engineering programs in 1990, 42 percent in 2000, and 44 percent in 2006 (National

NEW DIRECTIONS FOR INSTITUTIONAL RESEARCH, no. 152, Winter 2011 © Wiley Periodicals, Inc.
Published online in Wiley Online Library (wileyonlinelibrary.com) • DOI: 10.1002/ir.409

Science Foundation, 1995, 2008). These numbers illustrate positive gains for women in doctoral enrollment and graduation, but they mask a critical concern: completion for women in science and engineering fields is not occurring at the same rate as for their male counterparts.

Improving the graduation rate for women in science, engineering, and mathematics fields is not only a social justice and equity issue; it is an economic one. The United States continues to underproduce science and engineering graduates (Committee on Prospering in the Global Economy of the 21st Century, 2007) in an age where these degrees are in demand to meet global needs. The National Science Foundation (NSF) and the National Academy of Sciences have been charged with keeping the United States "at the leading edge of discovery in areas from astronomy to geology to zoology" (National Science Foundation, 2010). To do this, one of their main goals has been to increase the number of women in sciences and thus significantly boost the population of graduates able to fulfill the charge.

Attrition of Women in the Sciences

Research addressing doctoral women in the sciences has focused on increasing their representation in various fields (Clark and Corcoran, 1986; Etzkowitz, Kemelgor, and Uzzi, 2000; Margolis and Romero, 1998). However, retaining female students in the sciences has not been the focus of many of these studies. Lott, Gardener, and Powers (2009) conducted one of the few studies on doctoral student attrition that have implications for retaining women in science and engineering. The study, though not focused solely on women, showed that women were twice as likely to drop out as men after the seventh year in the program, but no differences prior to year seven were identified. The study also found that females in programs with higher female enrollment (greater than 30 percent) were less likely to drop out than from programs with higher male enrollment. The authors noted that having a critical mass was important for retaining female students through degree completion and thus suggest underrepresented students having other classmates who share similar characteristics are more likely to graduate.

Research studies of doctoral student retention show mixed results concerning whether there are significant differences in doctoral degree attainment between men and women, especially when all fields of study are considered. Some studies have shown that women are less likely to complete their degree (Bowen and Rudenstine, 1992; Stiles, 2003). These results are usually obtained when the studies use bivariate analysis or focus on a single field of study, such as education, physics, and so on. Studies that have used multivariate analyses or multiple academic fields show no significant differences between men and women in completing a doctoral degree (Ampaw and Jaeger, 2010; Nerad and Cerny 1991; Ott, Markewich, and Ochsner, 1984).

New Directions for Institutional Research • DOI: 10.1002/ir

These results can be interpreted in two ways. The first is since the statistics illustrate that fewer women complete a doctoral degree in science, engineering, and mathematics, but more women complete a doctoral degree in the humanities and social sciences, combining these programs cancels out the gender effect. The other explanation is that in multivariate analyses when including additional relevant variables in the model, the gender effect may be eliminated as a significant predictor because females are less likely to have this characteristic. For example, the literature has identified financial support as a significant factor in retention of doctoral students. Thus students who receive assistantships, especially research assistantships, are more likely to complete their degree (Ampaw and Jaeger, 2010; Ehrenberg and Mavros, 1995; Pyke and Sheridan, 1993; Gillingham, Seneca, and Taussig, 1991). However, if females are less likely to receive assistantships, then bivariate analysis would show that females are less likely to complete their degree since the analyses did not include assistantships. Consequently, key variables may not have been included in previous research that shows gender as a predictor of graduation. Further research should examine the conditional and interaction effects of important variables such as financial support by gender.

Finally, attrition research has not taken into account time-varying factors such as financial support and labor market information. Accounting for time variation in the variables includes the entire history of the variable instead of just one point in time, which offers a more vigorous analysis. For example, allowing assistantships to be a time-varying variable permits the analysis to account for all the semesters a student had a research assistantship, a teaching assistantship, or neither. Thus, the analysis is not limited to their assistantship type in their first year, final year, or a particular point in time. Lott, Gardener, and Powers (2009) do not address the time-varying explanatory variables in their model, and conducting logistic regression on a person-year dataset does not accurately estimate the effect of time-invariant variables in the model (age, gender). A person-year dataset is constructed by having a different line of data for each person for each given year he or she is present in the sample.

Conceptual Framework

Tinto's longitudinal model of doctoral persistence and human capital theory provide useful frameworks for this study. Tinto (1993) theorizes that doctoral persistence occurs when doctoral students are academically and socially integrated into their program, department, or field of study. Academic integration into a department is particularly important for doctoral students in science and engineering because successful completion of a dissertation often involves significant lab research with a research team that includes faculty. Tinto also theorizes that students' attributes and prior educational experiences will affect their goals and commitments, influencing their integration into a program. Financial aid will also shape their participation in the program (Tinto, 1993). Assistantship stipends are connected to faculty research endeavors and program or department teaching responsibilities. These connections to the department are more likely to foster successful academic and social integration.

This study incorporates human capital theory into Tinto's model to determine the effect of the labor market on doctoral degree completion. Human capital theory has been used by economists and higher education researchers to explain why individuals choose to pursue further education (Becker, 1962, 1964; Paulsen, 2001). This decision occurs when the individual expects the benefits of the pursuit to exceed the costs. The expected benefits and costs can be intrinsic as well as extrinsic. This study focuses on extrinsic benefits and costs because of their policy implications and the available data.

Description of Study

To address some of the shortcomings in the empirical literature to date, this chapter highlights the results from a study on students in the science, engineering, and mathematics fields that examined whether or not there was a gender effect on doctoral degree completion in these fields of study. The purpose of our study was to examine the effect of student characteristics, financial aid, and labor market conditions on the degree completion of doctoral students in SEM fields. The study used the interaction of these variables with gender to determine if any of these factors play a role in the degree completion of women in science and engineering.

The research questions that guided the study were:

- What are the effects of student characteristics, financial aid packages, and labor market conditions on the completion of a doctoral degree in SEM fields?
- What are the significant predictors of graduation for women in SEM fields?

We used three sources of data from a research-extensive institution in the southeast region of the United States to form the dataset. First, the institution's research office provided data on the students and departments. The Bureau of Labor Statistics furnished unemployment and weekly earnings information for the fields from data derived from its Current Population Study. The National Faculty Salary Survey by Discipline and Rank in Four Year Colleges and Universities, administered by the College and University Professional Association for Human Resources (CUPA-HR), generated the expected earnings information.

Information was obtained about doctoral students who enrolled at the institution in the four colleges between the academic years of 1994–95 and 1998–99. This allowed use of data from the most recent doctoral cohorts with at least ten years of information. The research used only five cohorts in order to reduce unmeasured variations in department and institution policies that would influence persistence between cohorts. Since the analytical period studied was from the years 1994 to 2008, there was enough variation in economic conditions to estimate their impact on retention.

The dependent variable for the study was whether the student had completed the doctoral degree by 2008. The institution had a ten-year maximum requirement for completion of the doctoral degree (except in special circumstances), and so this variable measured whether the student had completed the intended doctoral degree in ten years. Students who enrolled in the institution with the intent of obtaining a doctoral degree but graduated with a master's degree were considered to have dropped out of the doctoral program.

The independent variables fall into three categories: student characteristics, labor market information, and program information. Student characteristics used in the research are gender, age, race and citizenship, and having previously earned a master's degree. For the race and citizenship variable, international students were coded as one category and white students as another category. Because of the small samples of other races, they were combined into one minority category.

Doctoral students in the sample had varying enrollment patterns throughout their study period and as a result enrollment status was allowed to change between full-time and part-time throughout the study. Part-time status was determined by a student's enrollment each semester and classified as enrollment of fewer than nine credit hours and included in the estimation as a semester-varying dummy variable. The main sources of financial aid for doctoral students were assistantships, fellowships, and loans. The institutional research office did not collect information on the type of fellowship obtained by students but had information on all external grants paid to students through the institution, which generally tended to be fellowships with the highest sums coming from more prestigious fellowships. The monetary amounts were thus used as proxies for fellowships. Semester-varying financial aid measures were included: dummy coded variables for assistantships and natural log transformations of their monetary values for grants and loans.

For the department information, the student's college is included in the analysis as the size of the department as it relates to doctoral students per faculty using a continuous semester-varying measure. Lovitts (2001) showed a correlation between faculty retention rate and doctoral student attrition, so faculty turnover rate was also included in the estimations.

The study used three measures of the labor market. The first was the unemployment rate of skilled workers within the student's field of study, which provided a measure of employment availability. The second measure was the weekly wages that students would have earned by working instead of attending graduate school full-time. This was used as a proxy for the opportunity costs of attending a doctoral program. The final measure was salary for a new assistant professor within the field of study, which provided a measure of expected earnings after degree completion. Despite the fact that not all doctoral students aspire to a faculty career after graduation, the research used this measure for expected earnings because it was the best available indicator of what doctorate holders expect to earn with their degrees.

Analytical Method

This study used event history analysis, and more specifically a discrete-time hazard model as the main analytical tool to answer the research questions. The longitudinal nature of the study, the censored observations, and the time-varying nature of enrollment, financial aid, and the labor market lent themselves to use of event history analysis to answer the research questions (Cox, 1972). A discrete-time hazard model was chosen since the time unit used in the analysis (a semester) was discrete. This hazard model was also chosen since the data featured a large number of ties. In event history analysis, ties occur when two or more people experience the event at the same time, and within a given semester multiple students graduated in the sample, thus experiencing ties. To conduct the multivariate analyses, a Cox proportional model is used, which estimates the effect of each variable on the likelihood of completing the degree (Allison, 1982).

Summary of Findings

The descriptive results of the study showed that almost 10 percent more males in the study completed their doctoral degree in ten years than females (56 percent versus 47 percent respectively). In addition, females were more likely to be younger at the start of their doctoral program than males. Males in the study were concentrated in engineering, whereas females were concentrated in agricultural and life sciences.

Figure 6.1 shows the percentage of students with research assistant-ships in each semester by gender. The chart shows that more males received research assistantships than female students. At their highest point, 38 percent of male students received a research assistantship and only 16 percent of females received a research assistantship. The percentage of students receiving research assistantships declined the longer students stayed in their respective doctoral program. Figure 6.2 shows similar results for students receiving teaching assistantships by gender. The highest percentage of students receiving teaching assistantships occurred with the first year, and a higher percentage of male students received teaching assistantships than females.

NEW DIRECTIONS FOR INSTITUTIONAL RESEARCH • DOI: 10.1002/ir

Figure 6.1. Percentage of Students with Research Assistantships by Gender

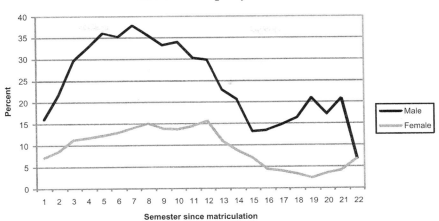

Figure 6.2. Percentage of Students with Teaching Assistantships by Gender

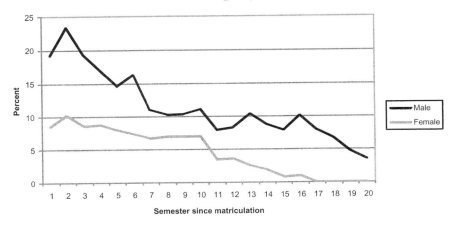

What Happens to Students in SEM over Time?

The results of event history analysis are summarized in this section. The first set of results presented is the bivariate analyses using graphs of the hazard functions for degree completion across time. A hazard function shows the probability of graduating in a particular semester for various groups of students. Figure 6.3 presents the hazard function of degree

Figure 6.3. Smoothed Hazard Estimates of Degree Completion

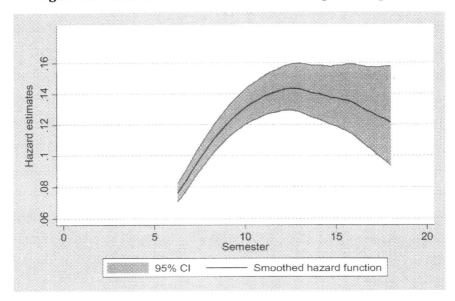

completion. The graph shows that students at the sixth semester have an 8 percent probability of graduating, if they are still enrolled. This probability increases steadily to the peak of 14 percent in the thirteenth semester and then declines. Thus students who do not complete their degree by the sixteenth semester reduce their likelihood of graduating for every additional semester they remained enrolled.

Figure 6.4 compares the hazard function of degree completion by gender. The two genders exhibit similar shapes in their hazard functions, but the curves are not equal, which implies that the time to degree completion for women is different from the one for men in science, engineering, and math. In other words, for all semesters males have a higher likelihood of graduating than females. At the peak semester, the twelfth after matriculation, males have a 15 percent probability of graduating if they are still enrolled while for females it is 13 percent.

Proportional Hazard Results. In this section we summarize the hazard ratio results of the proportional hazard regressions for completing the degree. Hazard ratios are similar to the odds ratios in logistic regression and report the effect independent variables have on the hazard rate of an event. The results confirm that females in science, engineering, and math have a lower likelihood of degree completion. Females are 32 percent less likely to graduate than their male colleagues.

Other variables in the model provide additional information to consider. For example, students who are older than thirty years on entry into

Figure 6.4. Smoothed Hazard Estimates of Degree Completion by Gender

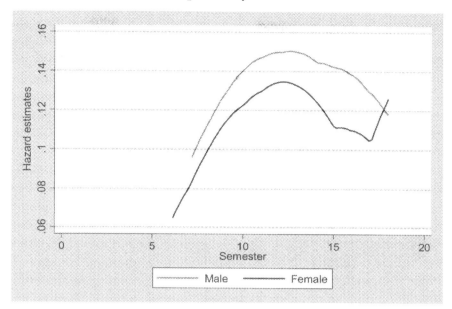

the doctoral program are less likely to complete the degree. The interaction of gender and age shows no significant difference, meaning that older females are equally as likely to not complete the degree as older male students. Furthermore, the results show that international students are more likely to complete the doctoral degree, but minority students in general do not have a significantly different likelihood of completion compared to their white counterparts. Readers should note that the numbers within the ethnicity groups were too small for analysis; thus, we grouped all minorities together, which dilutes the overall finding of ethnicity on the likelihood of completion.

Students with part-time status were more likely to complete the doctoral degree compared to students attending full-time, and a higher unemployment rate in the student's field increases the likelihood of degree completion. A 1 percent increase in (1) expected earnings in terms of assistant professor salaries and (2) forgone earnings increases the likelihood of degree completion by 3.8 percent and 2.7 percent, respectively. Students with research assistantships have a higher likelihood of completing their degree compared to students with teaching assistantships or no assistantships. Teaching assistantships did not significantly affect degree completion. Grants have no significant effect on the likelihood of degree completion; however, students with larger loans were less likely to complete the degree.

Faculty turnover had a significant impact on doctoral degree completion, but with the odds ratio close to 1:1 the impact appears to be very small. Students in the College of Engineering, Natural Resources, and Physical and Mathematical Sciences were less likely to complete their degree than the students in Agricultural and Life Sciences. However, the interaction of the college and gender show that females in the Colleges of Engineering and Natural Resources were more likely to complete their degree (see Table 6.1 for additional information).

Table 6.1. Results of Proportional Hazards for Graduating with a Doctoral Degree

Variable	Hazard Ratio	Standard Error
Student characteristics		
Female	0.676***	0.096
International student	1.920***	0.175
Minority student	0.775	0.120
Previous master's degree	1.004	0.082
Part-time	3.409***	0.414
Age (less than 25 years)		
25–30 years	1.103	0.13
Greater than 30 years	0.727***	0.103
Interaction of female and age		
25–30 years	0.997	0.204
Greater than 30 years	1.098	0.277
Labor market information		
Unemployment rate	1.058**	0.024
New assistant prof. salary (log)	3.811**	2.051
Earnings (log)	2.690**	1.287
Financial aid information		
Research assistantship	1.667***	0.203
Teaching assistantship	1.261	0.191
Grant (log)	0.981	0.015
Loan (log)	0.947***	0.018
Department information		
Students per faculty member	0.908	0.052
Faculty turnover	1.007**	0.004
College (Agriculture and Life Sciences)		
Engineering	0.413***	0.083
Natural Resources	0.592**	0.123
Physical and Mathematical Sciences	0.576***	0.097
Interaction of female and college		
Engineering	2.493***	0.531
Natural Resources	3.177***	1.304
Physical and Mathematical Sciences	1.328	0.297

Significance *** $p < 0.01$, ** $p < 0.05$

NEW DIRECTIONS FOR INSTITUTIONAL RESEARCH • DOI: 10.1002/ir

Discussion and Conclusions

The results from our study support that labor market conditions do affect doctoral degree completion for students in the SEM fields. Students in SEM fields with higher expected earnings will be more likely to complete their degree, and a higher unemployment rate reduces the likelihood that students will drop out from their program. These results are consistent with human capital theory and suggest that higher expected benefits from the degree will encourage completion. Differences related to gender were not identified.

Critical to the discussion of doctoral completion are funding opportunities. Research (Ampaw and Jaeger, 2010; Ehrenberg and Mavros, 1995) suggests that the type of funding doctoral students receive plays a role in whether they attain a degree. Similar results were identified in this study. Doctoral students with teaching assistantships are essential to many disciplines as they cover critical course sections for undergraduates. In addition, they provide funding for significant numbers of doctoral students. Yet these assistantships do not play a role in doctoral completion. Noteworthy to this discussion is the research assistantship position. Our findings revealed more males receiving research assistantships than females. In the semester in which the greatest number of research assistantships were attained, 38 percent of male students held such positions whereas only females held only 16 percent. These numbers are startling; students who hold research assistantships are 67 percent more likely to complete than those who do not. This finding may be uncovering what gender has been masking in previous research. The issue is not whether women graduate at a lower rate than men but that women are not attaining the essential research positions that would aid in their successful completion.

The departmental findings in this study provide additional insight into why students may or may not be completing doctoral degrees. Consistent with literature on the importance of mentoring in doctoral education (Ellis, 2001; Howard-Hamilton and others, 2009; Sweitzer, 2009), it is not surprising that faculty turnover negatively affects degree completion. Yet Ampaw and Jaeger's (2010) recent research, which included all doctoral fields, did not identify a consistent finding. The faculty-student relationship would appear to be even more critical in the science, math, and engineering departments where student funding is connected to faculty research; thus, if a faculty member leaves an institution, a student's funding may be eliminated.

Unlike Lott, Gardener, and Powers (2009), we did not observe the idea of critical mass in this study. Lott and colleagues noted that a critical mass of students who are like one another could explain when the completion rate in disciplines with fewer women was lower for women. In agriculture and life sciences, women represented 45.6 percent of the students (one could say they had a critical mass of women) but graduated at a lower rate than women in engineering and in natural resources who had smaller percentages of women (22 percent and 4 percent respectively). In engineering and natural resources women had a higher rate of completion than men. Although the number of women entering these programs is still far fewer than men, they are completing at a rate higher than men. This is also in contrast to recent Bell (2008) data that identify the completion rate for men at 9 percentage points higher than for women (65 percent men, 56 percent women). Given the efforts at this institution to support women in various programs in these departments, this finding is not completely unusual. This may also suggest that programming efforts through the NSF and other organizations may have positive effects on the completion rate of female doctoral students, although definitive results of programming efforts was beyond the scope of this study.

The effect of age at entry on doctoral degree completion has exhibited mixed results in previous studies. Studies by Lott, Gardener, and Powers (2009) and Nettles and Millet (2006) show that older students are more likely to complete the degree, while Abedi and Benkin (1987) found no effect of age on degree completion and Stiles (2003) showed that older students were disadvantaged at the later stages of their program. This study shows that despite women being on the average younger than their male counterparts and the results showing that age negatively affects degree completion for both groups (the older you get, the less likely you are to graduate), females have a lower completion rate. Thus the age at entry for female students is not a factor in their observed lower degree completion rate. It could be that for female students, marital status and number of children—which this study was unable to explore—are factors in their degree completion more so than age. Studies that have explored marital status showed married students are more likely to complete their degree (Lott, Gardener, and Powers, 2009; Lovitts, 2001), although this result has not been explored for women only. Additional research should include factors often related to gender issues, including children, partner, and family information.

Implications for Institutional Research

The results from our study yield implications for institutional research. One of the major findings was that the type of funding students receive to attend a doctoral program plays an important role in whether or not they complete their degree. Thus more research is needed that investigates why research assistantships are a critical element in the discussion of doctoral completion because few studies have addressed this issue. Although teaching assistantships are important to many institutions in providing students with funding and programs and departments with essential instructors for large numbers of undergraduate students, these assistantships do not aid in doctoral completion. On the other hand, students who hold research assistantships are 67 percent more likely to complete their degree than those who do not. Furthermore, in the semester in which the greatest number of research assistantships were obtained, 38 percent of male students had one compared to only 16 percent of females who held such positions. We suggest that addressing who is awarded highly sought-after research assistantships will also address who completes or fails to complete the doctoral degree. Thus institutional researchers should consider keeping track of this information to aid in student retention in doctoral programs. Programs, departments, and colleges can advertise research assistantship positions to diverse audiences and specifically target advertising venues that may have higher exposure to women, for example advertising to specific undergraduate programs that have higher female enrollments.

Institutional researchers can assist faculty and administrators within individual program areas or departments in determining the completion rate for their respective students. Since this was an institutional study, we suggest that the study be replicated at other institutions. Would researchers obtain similar results at smaller institutions with a high concentration of science, math, and engineering degrees? Would the effects of teaching and research assistantships be similar at other institutions? Institutional researchers and administrators within the financial aid and enrollment areas should seek specificity in the types of fellowships and research assistantships offered to students to see if differences among these types of aid exist. Finally, paying particular attention to the opportunities women have to seek and obtain these positions is critical to their future successful doctoral completion.

References

Abedi, J., and Benkin, E. "The Effects of Students' Academic, Financial, and Demographic Variables on Time to the Doctorate." *Research in Higher Education*, 1987, 27(1), 3–14.

Allison, P. D. "Discrete-Time Methods for the Analysis of Event Histories." *Sociological Methodology*, 1982, 13, 61–98.

Ampaw, F. D., and Jaeger, A. J. "Completing the Three Stages of Doctoral Education: An Event History Analysis." Paper presented at the Association for the Study of Higher Education Conference, Indianapolis, Ind., Nov. 2010.

Becker, G. S. "Investment in Human Capital: A Theoretical Analysis." *The Journal of Political Economy*, 1962, 70(S5), 9–49.

Becker, G. S. *Human Capital: A Theoretical and Empirical Analysis, with Special reference to Education*. New York: National Bureau of Economic Research, 1964.

Bell, N. *Graduate Enrollment and Degrees: 1997 to 2007*. Washington, D.C.: Council of Graduate Schools, 2008.

Bowen, W. G., and Rudenstine, N. L. *In Pursuit of the Ph.D.* Princeton, N.J.: Princeton University Press, 1992.

Clark, S. M., and Corcoran, M. "Perspectives on the Professional Socialization of Women Faculty: A Case of Accumulative Disadvantage?" *Journal of Higher Education*, 1986, 57, 20–43.

Committee on Prospering in the Global Economy of the 21st Century. *Rising Above the Gathering Storm: Energizing and Employing America for a Brighter Economic Future.* Washington, D.C.: National Academies Press, 2007.

Cox, D. R. "Regression Models and Life Tables (with Discussion)." *Journal of the Royal Statistical Society, Series B*, 1972, *34*(2), 187–220.

Ehrenberg, R. G., and Mavros, P. "Do Doctoral Students' Financial Support Patterns Affect Their Times-to-Degree and Completion Probabilities?" *Journal of Human Resources*, 1995, *30*(3), 581–609.

Ellis, E. M. "The Impact of Race and Gender on Graduate School Socialization, Satisfaction with Doctoral Study, and Commitment to Degree Completion." *Western Journal of Black Studies*, 2001, *25*(1), 30–45.

Etzkowitz, H., Kemelgor, C., and Uzzi, B. *Athena Unbound: The Advancement of Women in Science and Technology.* Cambridge: Cambridge University Press, 2000.

Gillingham, L., Seneca, J. J., and Taussig, M. K. "The Determinants of Progress to the Doctoral Degree." *Research in Higher Education*, 1991, *32*(4), 449–468.

Howard-Hamilton, M. F., and others. (eds.). *Standing on the Outside Looking In: Underrepresented Students' Experiences in Advanced Degree Programs.* Sterling, Va.: Stylus, 2009.

Lott, J. L., Gardener, S., and Powers, D. A. "Doctoral Student Attrition in the STEM Fields: An Exploratory Event History Analysis." *Journal of College Student Retention*, 2009, *11*(2) 247–266.

Lovitts, B. E. *Leaving the Ivory Tower: The Causes and Consequences of Departure from Doctoral Study.* Lanham, Md.: Rowman & Littlefield, 2001.

Margolis, E., and Romero, M. "'The Department Is Very Male, Very White, Very Old, and Very Conservative': The Functioning of the Hidden Curriculum in Graduate Sociology Departments." *Harvard Educational Review*, 1998, *68*(1), 1–32.

National Science Foundation (NSF). "Graduate Students and Postdoctorates in Science and Engineering: Fall 1994." 1995. Retrieved November 4, 2008, from http://www.nsf.gov/statistics/gradpostdoc/94supp/.

National Science Foundation (NSF). "*Graduate Students and Postdoctorates in Science and Engineering: Fall 2007.*" 2008. Retrieved November 4, 2008, from http://nsf.gov/statistics/gradpostdoc/.

National Science Foundation (NSF). "About the National Science Foundation." 2010. Retrieved December 27, 2010, from http://nsf.gov/about/.

Nerad, M., and Cerny, J. "From Facts to Action: Expanding the Educational Role of the Graduate Division." *CGS Communicator; Special Edition*, 1–12, 1991.

Nettles, M. T., and Millett, C. M. *Three Magic Letters: Getting to Ph.D.* Baltimore, Md.: Johns Hopkins University Press, 2006.

Ott, M. D., Markewich, T. S., and Ochsner, N. L. "Logit Analysis of Graduate Student Retention." *Research in Higher Education*, 1984, *21*, 439–460.

Paulsen, M. B. "The Economics of Human Capital and Investment in Higher Education." In M. B. Paulsen and J. C. Smart (eds.), *The Finance of Higher Education: Theory, Research, Policy and Practice.* New York: Agathon, 2001.

Pyke, S. W., and Sheridan, P. M. "Logistic Regression Analysis of Graduate Student Retention." *Canadian Journal of Higher Education*, 1993, *23*, 44–64.

Stiles, J. E. *The Hazard of Success: A Longitudinal Study of Time to Degree Among Doctoral Students Using Discrete-Time Survival Analysis.* Unpublished doctoral dissertation, Harvard University, School of Education, 2003.

Sweitzer, V. "Toward a Theory of Doctoral Student Professional Identity Development: A Developmental Theory Approach." *Journal of Higher Education*, 2009, *80*(1), 1–33.

Tinto, V. *Leaving College: Rethinking the Causes and Cures of Student Attrition (2nd ed.).* Chicago: University of Chicago Press, 1993.

FRIM D. AMPAW *is an assistant professor of educational leadership at Central Michigan University. Her research investigates the retention of doctoral, female, and minority students in the STEM fields.*

AUDREY J. JAEGER *is an associate professor of higher education and co-executive director of the National Initiative for Leadership and Institutional Effectiveness at North Carolina State University. Her research examines the roles of current and future faculty with emphasis on STEM faculty and part-time faculty as well as exploring how cultural and social identity influences the career choices of doctoral students of color.*

NEW DIRECTIONS FOR INSTITUTIONAL RESEARCH • DOI: 10.1002/ir

Treating women as a homogeneous group obscures important racial and ethnic differences among women in STEM. This chapter focuses on the experiences of women of color in science and engineering and highlights the importance of addressing intersecting identities among women.

Women of Color in Science, Technology, Engineering, and Mathematics (STEM)

Dawn R. Johnson

Scholars have theorized and examined women's underrepresentation in science, technology, engineering and mathematics (STEM) fields for well over thirty years (Clewell and Campbell, 2002). However, much of this research has paid little attention to issues of racial and ethnic diversity among women, suggesting that all women have the same experiences in STEM (Clewell and Ginorio, 1996; Hanson, 2004). Women of color were excluded from research designs, or when they were included their numbers were too small for any meaningful analysis (Clewell and Ginorio, 1996). Some researchers did not describe the racial and ethnic composition of their samples, even if only to note that all of the participants were white women, or simply chose not to examine racial and ethnic differences in the design of the study (Atwater, 2000). Ignoring race and ethnicity obscures important dimensions of women's experiences in STEM and fosters the notion of a universal gender experience among women, without considering the differential experiences of women of color or the effects of racial privilege for white women (Atwater, 2000; Collins, 1999; Hanson, 2004). Women of color should not be further marginalized in a body of research that, in theory, is about transforming women's underrepresentation in male-dominated academic disciplines and career fields.

There are many calls for inclusion of women of color in STEM research and policy development in conferences and symposia (Malcom, Hall, and Brown, 1976; Ong, 2010), published literature reviews (Clewell and Anderson, 1991; Ong, Wright, Espinosa, and Orfield, 2011), and scholarly critiques (Atwater, 2000; Clewell and Ginorio, 1996; Collins,

New Directions for Institutional Research, no. 152, Winter 2011 © Wiley Periodicals, Inc.
Published online in Wiley Online Library (wileyonlinelibrary.com) • DOI: 10.1002/ir.410

1999). A body of research on women of color in STEM has emerged to include research on African American (Hanson, 2009, 2004; Perna and others, 2009), Asian American (Chinn, 2002, 1999), and Latina (Brown, 2008) women, as well as women from other racial and ethnic groups (Carlone and Johnson, 2007; Ong, 2005; Trenor and others, 2008). Much of the published research used qualitative methods (Foor, Walden, and Trytten, 2007; Johnson, 2007; Tate and Linn, 2005), while others used quantitative analyses on large datasets (Brown, 1995; Espinosa, 2011; Hanson, 2009; Johnson, forthcoming; Nelson and Brammer, 2010) or employed a mixed-methods approach (Trenor and others, 2008). There were also several studies that included gender and race or ethnicity as variables to analyze the experiences of women of color (Bonous-Hammarth, 2000; Grandy, 1997, 1998; Huang, Taddese, Walter, and Peng, 2000; Leslie, McClure, and Oaxaca, 1998; Malone and Barabino, 2009; Smyth and McArdle, 2004). The majority of the research on women of color in STEM has focused on undergraduate women, with fewer studies examining graduate students or faculty (Ong, 2010).

In 2008, women of color accounted for 34 percent of all female undergraduate students and earned 26 percent of the bachelor's degrees awarded to women across all fields of study (National Science Foundation, 2011). These demographics suggest the need for inclusive scholarship on women in STEM because "it is no longer feasible, or appropriate, or just to conduct research solely on white subjects and use the findings of this research to make policy decisions for the entire populace" (Clewell and Ginorio, 1996, p. 216). To that end, this chapter provides an overview of how women of color are represented in higher education as undergraduate and graduate students and faculty. The unique challenges of race and ethnicity for women of color in STEM are discussed, as well as suggestions for institutional research practices that can build on the knowledge of women of color in STEM.

Definitions of STEM Fields and Women of Color

There are variations in the identification of particular academic disciplines as "STEM." For example, the National Science Foundation (NSF) uses the term *science and engineering* to include agricultural, biological, and computer sciences; atmospheric, earth, and ocean sciences; mathematics and statistics; astronomy; chemistry; physics; aerospace, chemical, civil, electrical, industrial, materials, and mechanical engineering; social sciences, and psychology (National Science Foundation, 2011). Using this definition, the NSF reported that women earned 50 percent of the bachelor's degrees in science and engineering fields in 2008; however, when the social science and psychology fields are excluded, the data show that women earned 39 percent of science and engineering bachelor degrees

NEW DIRECTIONS FOR INSTITUTIONAL RESEARCH • DOI: 10.1002/ir

(National Science Foundation, 2011). A report authored by Chen and Weko (2009) using data from the National Center of Education Statistics defines *STEM fields* as including the natural sciences (physical, biological, and agricultural sciences), engineering and engineering technologies, computer and information sciences, and mathematics. Chen and Weko (2009) noted their definition of STEM excludes social sciences and psychology because various state and federal legislative actions regarding STEM access and education do not include these fields. The STEM data from the NSF (2011) reported in this chapter do not include social science and psychology fields because these areas of study are heavily populated by women.

In discussing women of color in the context of STEM fields, most research and policy reports focus on black/African American, Latina/Hispanic, and Native American/American Indian women because these racial and ethnic groups are underrepresented in STEM fields relative to their representation in the overall population in the United States (Clewell and Anderson, 1991; Malcom, Hall, and Brown, 1976; National Science Foundation, 2011; Ong, Wright, Espinosa, and Orfield, 2011). Women from those racial and ethnic groups are referred to as "under-represented women" (Towns, 2010), "underrepresented minority women" (Nelson and Brammer, 2010), or "women from underrepresented groups" (Hill, Corbett, and St. Rose, 2010). Asian Pacific American women may be included in research on women of color in STEM (Carlone and Johnson, 2007; Ong, Wright, Espinosa, and Orfield, 2011), but more often they are excluded because they are not underrepresented in STEM fields relative to their representation in the U.S. population (National Science Foundation, 2011). However, this exclusion obscures the different rates and patterns of STEM participation among the ethnic groups designated as Asian Pacific American (Lee, 1997), and it does not account for the racial discrimination faced by Asian Pacific American women (Chinn, 2002, 1999) or for their earning fewer STEM degrees than their male counterparts (Hill, Corbett, and St. Rose, 2010). As the number of individuals identifying as multiracial increases, STEM policymakers must consider how to include this group of women (Ong, Wright, Espinosa, and Orfield, 2011) because scholars already include multiracial women in their research on women of color (Carlone and Johnson, 2007; Foor, Walden, and Trytten, 2007; Tate and Linn, 2005). This chapter defines *women of color* as black/African American, Latina/Hispanic, Native American/American Indian, Asian Pacific American, and multiracial women, recognizing that although women of color may share the common experience of racial oppression and discrimination, each racial and ethnic minority group also has unique social, economic, and political histories that contribute to their experiences of oppression and discrimination in the U.S. educational system (Andersen and Collins, 2001).

NEW DIRECTIONS FOR INSTITUTIONAL RESEARCH • DOI: 10.1002/ir

Patterns of Representation Among Students and Faculty

Data indicate that as undergraduate students, women of color expressed greater intention to major in STEM at the start of their freshman year of college than white women (National Science Foundation, 2011; Smyth and McArdle, 2004). In 2008, 13.4 percent of white female first-year students indicated their interest in majoring in STEM, compared to 29 percent of Asian Pacific American women and 16 percent of women from each of the underrepresented racial and ethnic groups (National Science Foundation, 2011). Enrollment data by undergraduate major are available only for engineering fields; these data indicate that women's enrollment in undergraduate engineering majors peaked at close to 20 percent in 1999 and hit a ten-year low in 2006 at 17 percent (National Science Foundation, 2009), with only a slight increase to 17.5 in 2008 (National Science Foundation, 2011). The racial and ethnic composition of women enrolled in engineering majors in 2008 (among all students) included 10.7 percent white, 2.2 percent Asian, 2 percent Hispanic, 1.4 percent black, 1 percent foreign national, and less than 1 percent Native American women (National Science Foundation, 2011). As highlighted in the first chapter of this volume, factors related to choosing a STEM major for women of color include high school math and science preparation, mother's level of education, expectations from parents of completing college, having a parent who works in a STEM field, and having a positive attitude toward science (Hanson, 2004; Huang, Taddese, Walter, and Peng, 2000; Leslie, McClure, and Oaxaca, 1999).

Women earned 38 percent of the STEM baccalaureate degrees (excluding psychology and social science fields) in the United States in 2008, including 18 percent in engineering and computer science fields (National Science Foundation, 2011). Among all students earning bachelor's degrees in STEM in 2008, women of color from underrepresented groups earned 6.1 percent, Asian Pacific American women earned 4.8 percent, white women earned 23.5 percent, women identified as "other" or race unknown earned 2 percent, and women identified as temporary residents earned 1.4 percent (National Science Foundation, 2011). Women from underrepresented groups had the lowest rate of STEM persistence among all students in STEM (Bonous-Hammarth, 2000; Grandy, 1998; Smyth and McArdle, 2004). Factors related to STEM persistence for women of color include SAT math scores, grades from high school, parental level of education and employment in a STEM field, and having plans to attend graduate school in a STEM field (Bonous-Hammarth, 2000; Huang, Taddese, Walter, and Peng, 2000; Leslie, McClure, and Oaxaca, 1999; Smyth and McArdle, 2004).

At the graduate level, women made up 34.4 percent of the graduate students enrolled in STEM fields in 2008, with 14.8 percent white women,

1.6 percent black/African American women, 2.6 percent Asian Pacific Islander women, 1.5 percent Hispanic/Latina, and less than 1 percent American Indian women. Women with temporary resident status were 11.8 percent and women indicating their race as "other" or whose race or ethnicity was unknown were 2 percent of graduate students in STEM fields (National Science Foundation, 2011). Nearly 33 percent of master's degrees in STEM fields were earned by women in 2008; white women earned 13.6 percent, black/African American women earned 1.7 percent, Asian Pacific Islander women earned 3.4 percent, Hispanic/Latina earned 1.3 percent, American Indian women earned less than 1 percent, women identified as "other" or whose race or ethnicity was unknown earned 2.2 percent, and women with temporary resident status earned 10.4 percent (National Science Foundation, 2011). Among the doctorates conferred in STEM fields in 2008, 37.4 percent were earned by women; 25.6 percent were white, 4.7 percent were Asian Pacific Islander, 2.1 percent were black/African American, 1.8 were Hispanic/Latina, fewer than 1 percent were American Indian, and 3 percent were women identified as "other" (National Science Foundation, 2011).

An examination of women's representation among STEM faculty across discipline and rank indicates that women were well represented in life sciences fields, making up 32.3 percent of associate and assistant professors and 22 percent of full professors in 2008. Women were less represented in computer science, engineering, and the physical sciences, with engineering having the fewest women on faculty (17.5 percent associate and assistant professors and 5 percent of full professors in 2008; National Science Foundation, 2011). However, the portrait for women of color faculty in STEM is quite bleak. Nelson and Brammer's 2010 report on faculty diversity in science and engineering at research universities indicated that out of 14,400 tenured or tenure-track faculty in the top fifty departments in the United States, 1,678 faculty were women and 88 faculty were women from underrepresented racial or ethnic groups. Towns (2010) made a simple yet important observation: that increasing the number of women of color on STEM faculty requires supporting them at the undergraduate and graduate levels.

The "Double Bind" for Women of Color in STEM

In describing the experiences of women of color in STEM, Malcom, Hall, and Brown (1976) used the term *double bind* to convey the idea that women of color as scientists experience oppression and discrimination based on their race or ethnicity and gender, resulting in women of color being the least recognized and valued, and the most invisible and marginalized, among underrepresented groups in STEM (Malcom, Hall, and Brown, 1976). Evidence of the double bind exists in detailed accounts from women of color about numerous instances of racism and sexism

NEW DIRECTIONS FOR INSTITUTIONAL RESEARCH • DOI: 10.1002/ir

encountered in STEM educational environments (Foor, Walden, and Tryt-ten, 2007; Johnson, 2007; Malone and Barabino, 2009; Ong, 2005; Ong, Wright, Espinosa, and Orfield, 2011). As a result of being in both the racial and ethnic and the gender minority groups, women of color experi-enced isolation from peers and a lack of support from faculty, while vari-ous sociocultural factors influenced the way women of color negotiated the culture, values, and practices in the sciences. However, even though some women of color paid the price for the double bind by leaving their STEM educations and aspirations, others found agency, resilience, and support in the midst of the racial and gender oppression (Brown, 2008; Johnson, 2005; Ong, 2005; Tate and Linn, 2005).

Exclusion, Isolation, and Lack of Belonging. Women of color in STEM reported feeling excluded in the academic environment when they were avoided by white students in choosing where to sit in a classroom, and selecting partners for group assignments, laboratory work, and infor-mal study groups (Johnson, 2007; Malone and Barabino, 2009; Ong, 2005; Tate and Linn, 2005). Women of color were isolated as "the only one" in the classroom or laboratory (Malone and Barabino, 2009) and when they were not included in informal networking and socializing among students and faculty where useful information was shared about classroom and lab-oratory work, as well as scholarship and research opportunities (Foor, Walden, and Trytten, 2007; Malone and Barabino, 2009; Tate and Linn, 2005). The lack of racial and ethnic diversity in STEM meant that women of color from underrepresented groups had fewer racial and ethnic group peers to provide critical academic and social support (Tate and Linn, 2005). Less-than-positive perceptions of the broader racial climate on campus negatively contributed to overall sense of belonging for women of color in STEM (Johnson, forthcoming), and the extent to which women of color felt their institution was committed to racial and ethnic diversity on campus and in STEM fields (Malone and Barabino, 2009).

Because much of the work in STEM is done in teams, the isolation had a significant impact on the extent to which women of color experi-enced belonging to their STEM major or department (Foor, Walden, and Trytten, 2007). This sense of belonging affected how women of color developed and maintained their identity as scientists (Carlone and John-son, 2007; Ong, 2005; Tate and Linn, 2005). Women of color who were recognized by peers and faculty for their contributions and expertise had positive identities as scientists; those whose work went unnoticed and unacknowledged had difficulty establishing their science identity within their academic communities (Carlone and Johnson, 2007; Malone and Barabino, 2009; Ong, 2005; Tate and Linn, 2005).

Lack of Faculty Support. STEM learning environments for under-graduate students typically feature large lecture classes at the introductory level in which students compete with each other during class for the atten-tion and recognition of the instructor (Johnson, 2007; Seymour and

Hewitt, 1997). These learning environments were not supportive for many students (Seymour and Hewitt, 1997), especially for women of color, who were discouraged by faculty from continuing in their major when seeking help with difficult course material; faced with the racial stereotypes faculty had about their academic abilities; and ignored by faculty during classroom discussions, research meetings, or laboratory work (Brown, 2008; Foor, Walden, and Trytten, 2007; Malone and Barabino, 2009; Ong, 2005). Women of color were less likely to be highly recommended by faculty for fellowship opportunities (Brown, 1995) and received little guidance and support from faculty as they completed their research and looked for fellowship, grant, and internship opportunities (Malone and Barabino, 2009). Faculty in STEM often actively discouraged discussion of diversity issues in class, which left women of color feeling silenced and ignored (Johnson, 2007).

Social and Cultural Influences. Science culture is identified with notions of whiteness and masculinity, even though science is portrayed as being objective and neutral regarding issues of racial, ethnic, cultural, and gender differences (Johnson, 2007; Seymour and Hewitt, 1997). The culture of science is a meritocracy that is seen as competitive, difficult, and intellectually superior to other academic and professional fields because of the required technical and scientific expertise (Burack and Franks, 2006; Johnson, 2007; Seymour and Hewitt, 1997). These values stood in sharp contrast to the cultural values for women of color who strongly identified with their racial, ethnic and cultural backgrounds, and this made it difficult to navigate STEM environments. Cultural conflicts existed in performing laboratory procedures (for example, dissecting animals), asserting oneself in classroom and laboratory spaces, and competing with peers for grades, correct answers, and recognition from faculty (Carlone and Johnson, 2007; Foor, Walden, and Trytten, 2007; Johnson, 2007; Ong, 2005; Seymour and Hewitt, 1997).

Managing multiple demands from the role of student and from family responsibilities also shaped how women of color experienced STEM. Some Latinas described conflict between the expectations of fulfilling traditional gender roles within their families and the desire to pursue their college education and STEM major (Brown, 2008; Trenor and others, 2008). Asian Pacific American women also described meeting familial responsibilities and roles (Chinn, 2002) and family expectations of choosing a career in science and engineering (Trenor and others, 2008). The socioeconomic realities of many women of color necessitated working long hours to pay for college and support their families, and commuting from home rather than living on or near campus, which in turn affected their ability to participate in study groups, internships and other research experiences (Foor, Walden, and Trytten, 2007; Trenor and others, 2008).

Agency, Resilience, and Sources of Support. Even though women of color reported many challenges in their STEM experiences, agency and resilience were found among those who persisted to degree completion. Campus resources, including support groups, learning communities, undergraduate research programs, and student organizations geared toward students of color or women, were helpful for women of color (Brown, 2008; Espinosa, 2011; Johnson, 2005; Tate and Linn, 2005; Trenor and others, 2008). Academic resources such as course-related enrichment seminars, tutoring, study skills courses, and academically supportive residence halls were also important for women of color (Brown, 2008; Johnson, 2005, forthcoming; Tate and Linn, 2005; Trenor and others, 2008). Family (Hanson, 2004, 2009; Foor, Walden, and Trytten, 2007; Trenor and others, 2008), racial and ethnic peer groups outside of STEM majors (Tate and Linn, 2005), and mentoring experiences from alumni, faculty, and upper-level students were important forms of social support for women of color (Brown, 2008; Johnson, 2005; Tate and Linn, 2005; Trenor and others, 2008).

A strong sense of racial or ethnic identity gave many women of color the resilience and agency to cope with the stereotypes and marginalization they experienced in STEM (Brown, 2008; Malone and Barabino, 2009; Tate and Linn, 2005). Some women of color used their racial or ethnic and gender identity to manipulate stereotypes and gain visibility and credibility among peers and faculty (Ong, 2005), while other women of color used culturally relevant and humanistic goals for pursuing a career in science (e.g., giving back to their community, being a role model for girls of color, helping their family, wanting to make a difference in society) as tools for persisting in STEM (Espinosa, 2011; Trenor and others, 2008) and creating their science identity (Carlone and Johnson, 2007).

Implications for Developing Inclusive Research Practices

Women of color account for a growing number of undergraduate students and bachelor's degree recipients supplying the pipeline for careers in STEM fields (National Science Foundation, 2011). Understanding access and retention for women of color in STEM is critical for the development of the diverse scientific workforce necessary for addressing national and global issues (Ong, Wright, Espinosa, and Orfield, 2011). Increasing knowledge about women of color in STEM areas requires inclusive research practices, such as developing research designs that examine racial and ethnic differences, describing the racial and ethnic composition of samples, and considering issues of race and ethnicity in interpretation and discussion of the results, along with acknowledgment of the limitations associated with racially homogeneous samples. In studying students of color in STEM, gender differences should be analyzed and discussed, and multiracial women should be included in research samples given their increased representation on many campuses. Finally, research practice can incorporate multiple forms of data, such as using institutional records to augment data from individual interviews, focus groups, or surveys; and examining the experiences of women of color at the undergraduate, graduate, and faculty levels to gain multiple perspectives on the STEM environment and inform institutional policy development and practice.

Developing inclusive research practices also means addressing issues relevant to women of color in STEM. Further research is needed on the racial climate in STEM departments and classrooms and how this climate contributes to the decisions of women of color to stay or leave STEM fields at the undergraduate, graduate, and faculty levels. More understanding is needed about how women of color develop their racial and ethnic identities in STEM contexts and how these identities help women of color construct science identities and develop resilience in STEM. Finally, more research is needed about the teaching and advising practices of STEM faculty who have created supportive learning environments for women of color and how these pedagogies can be used to transform STEM educational practice. Taken together, these directions for inclusive research practices allow development of STEM educational policies and programs that take into account the needs of women of color, and bring complexity and depth to the scholarship on women in STEM by addressing the diversity of women's experiences in these fields.

References

Andersen, M. L., and Collins, P. H. *Race, Class, and Gender: An Anthology* (4th ed.). Belmont, Calif.: Wadsworth, 2001.

Atwater, M. M. "Females in Science Education: White Is the Norm and Class, Language, Lifestyle, and Religion Are Nonissues." *Journal of Research in Science Teaching*, 2000, 37(4), 386–387.

Bonous-Hammarth, M. "Pathways to Success: Affirming Opportunities for Science, Mathematics, and Engineering Majors." *Journal of Negro Education*, 2000, 69(1/2), 92–111.

Brown, S. V. "Testing the Double Bind Hypothesis: Faculty Recommendations of Minority Women Fellowship Applicants." *Journal of Women and Minorities in Science and Engineering*, 1995, 2(4), 207–225.

Brown, S. W. "The Gender Differences: Hispanic Females and Males Majoring in Science or Engineering." *Journal of Women and Minorities in Science and Engineering*, 2008, 14(2), 205–223.

Burack, C., and Franks, S. E. "Telling Stories About Engineering: Group Dynamics and Resistance to Diversity." In J. M. Bystydzienski and S. R. Bird (eds.), *Removing Barriers: Women in Academic Science, Technology, Engineering and Mathematics*. Bloomington: Indiana University Press, 2006.

Carlone, H. B., and Johnson, A. "Understanding the Science Experiences of Successful Women of Color: Science Identity as an Analytic Lens." *Journal of Research in Science Teaching*, 2007, 44(8), 1187–1218.

Chen, X., and Weko, T. *Students Who Study Science, Technology, Engineering, and Mathematics (STEM) in Post-Secondary Education*. Washington, D.C.: National Center for Education Statistics, 2009. Retrieved Jan. 30, 2010, from http://nces.ed.gov/pubs2009/2009161.pdf.

NEW DIRECTIONS FOR INSTITUTIONAL RESEARCH • DOI: 10.1002/ir

Chinn, P.W.U. "Multiple Worlds/Mismatched Meanings: Barriers to Minority Women Engineers." *Journal of Research in Science Teaching*, 1999, *36*(6), 621–636.

Chinn, P.W.U. "Asian and Pacific Islander Women Scientists and Engineers: A Narrative Exploration of Model Minority, Gender, and Racial Stereotypes." *Journal of Research in Science Teaching*, 2002, *39*(4), 302–323.

Clewell, B. C., and Anderson, B. *Women of Color in Mathematics, Science, and Engineering: A Review of the Literature*. Washington, D.C.: Center for Women Policy Studies, 1991.

Clewell, B. C., and Campbell, P. B. "Taking Stock: Where We've Been, Where We Are, Where We're Going." *Journal of Women and Minorities in Science and Engineering*, 2002, *8*(3/4), 255–284.

Clewell, B. C., and Ginorio, A. B. "Examining Women's Progress in the Sciences from the Perspective of Diversity." In C. Davis, A. B. Ginorio, C. S. Hollenshead, B. B. Lazarus, P. M. Rayman, and Associates (eds.), *The Equity Equation: Fostering the Advancement of Women in the Sciences, Mathematics, and Engineering*. San Francisco: Jossey-Bass, 1996.

Collins, P. H. "Moving Beyond Gender: Intersectionality and Scientific Knowledge." In M. M. Ferree, J. Lorber, and B. B. Hess (eds.), *Revisioning Gender*. Thousand Oaks, Calif.: SAGE, 1999.

Espinosa, L. L. "Pipelines and Pathways: Women of Color in Undergraduate STEM Majors and the College Experiences That Contribute to Persistence." *Harvard Educational Review*, 2011, *81*(2), 209–240.

Foor, C. E., Walden, S. E., and Trytten, D. A. "'I Wish That I Belonged More in This Whole Engineering Group': Achieving Individual Diversity." *Journal of Engineering Education*, 2007, *96*(2), 103–115.

Grandy, J. "Gender and Ethnic Differences in the Experiences, Achievements, and Expectations of Science and Engineering Majors." *Journal of Women and Minorities in Science and Engineering*, 1997, *3*(3), 119–143.

Grandy, J. "Persistence in Science of High-Ability Minority Students: Results of a Longitudinal Study." *Journal of Higher Education*, 1998, *69*(6), 589–620.

Hanson, S. L. "African American Women in Science: Experiences from High School Through the Post-Secondary Years and Beyond." *NWSA Journal*, 2004, *16*(1), 96–115.

Hanson, S. L. *Swimming Against the Tide: African American Girls in Science Education*. Philadelphia, Pa.: Temple University Press, 2009.

Hill, C., Corbett, C., and St. Rose, A. *Why So Few? Women in Science, Technology, Engineering, and Mathematics*. Washington, D.C.: American Association of University Women, 2010.

Huang, G., Taddese, N., Walter, E., and Peng, S. S. *Entry and Persistence of Women and Minorities in College Science and Engineering Education*. Washington, D.C.: National Center for Education Statistics, 2000. Retrieved Mar. 30, 2010, from http://nces.ed.gov/pubs2000/2000601.pdf.

Johnson, A. "Policy Implications of Supporting Women of Color in the Sciences." *Women, Politics, and Policy*, 2005, *27*(3/4), 135–150.

Johnson, A. "Unintended Consequences: How Science Professors Discourage Women of Color." *Science Education*, 2007, *91*(5), 805–821.

Johnson, D. R. "Campus Racial Climate Perceptions and Overall Sense of Belonging Among Racially Diverse Women in STEM Majors." *Journal of College Student Development*, forthcoming.

Lee, O. "Diversity and Equity for Asian American Students in Science Education." *Science Education*, 1997, *81*(1), 107–122.

Leslie, L. L., McClure, G. T., and Oaxaca, R. L. "Women and Minorities in Science and Engineering: A Life Sequence Analysis." *Journal of Higher Education*, 1998, *69*(3), 239–276.

Malcom, S. M., Hall, P. Q., and Brown, J. W. *The Double Bind: The Price of Being a Minority Woman in Science.* Washington, D.C.: American Association for the Advancement of Science, 1976.

Malone, K. R., and Barabino, G. "Narrations of Race in STEM Research Settings: Identity Formation and Its Discontents." *Science Education*, 2009, *93*(3), 485–510.

National Science Foundation (NSF). *Women, Minorities, and Persons with Disabilities in Science and Engineering: 2009.* Arlington, Va.: National Science Foundation, 2009. Retrieved, October 10, 2011, from http://www.nsf.gov/statistics/wmpd/pdf/nsf09305. pdf.

National Science Foundation (NSF). *Women, Minorities, and Persons with Disabilities in Science and Engineering: 2011.* Arlington, Va.: National Science Foundation, 2011. Retrieved, October 10, 2011, from http://www.nsf.gov/statistics/wmpd/start.cfm.

Nelson, D. J., and Brammer, C. N. *A National Analysis of Minorities in Science and Engineering Faculties at Research Universities.* Norman: University of Oklahoma, 2010. Retrieved January 17, 2011, from http://chem.ou.edu/~djn/diversity/Faculty_Tables_FY07/FinalReport07.html.

Ong, M. "Body Projects of Young Women of Color in Physics: Intersections of Gender, Race, and Science." *Social Problems*, 2005, *52*(4), 593–617.

Ong, M. *The Mini-Symposium on Women of Color in Science, Technology, Engineering, and Mathematics (STEM): A Summary of Events, Findings, and Suggestions.* Arlington, Va.: National Science Foundation, 2010. Retrieved September 10, 2010, from http://www.nsf.gov/od/oia/activities/ceose/reports/TERC_mini_symp_rprt_hires.pdf.

Ong, M., Wright, C., Espinosa, L. L., and Orfield, G. "Inside the Double Bind: A Synthesis of Empirical Research on Undergraduate and Graduate Women of Color in Science, Technology, Engineering, and Mathematics." *Harvard Educational Review*, 2011, *81*(2), 172–208.

Perna, L., and others. "The Contribution of HBCU's to the Preparation of African American Women for STEM Careers: A Case Study." *Research in Higher Education*, 2009, *50*(1), 1–23.

Seymour, E., and Hewitt, N. M. *Talking About Leaving: Why Undergraduates Leave the Sciences.* Boulder, Colo.: Westview Press, 1997.

Smyth, F. L., and McArdle, J. J. "Ethnic and Gender Differences in Science Graduation at Selective Colleges with Implications for Admission Policy and College Choice." *Research in Higher Education*, 2004, *45*(4), 353–381.

Tate, E. D., and Linn, M. C. "How Does Identity Shape the Experiences of Women of Color Engineering Students?" *Journal of Science Education and Technology*, 2005, *14*(5/6), 483–493.

Towns, M. H. "Where Are the Women of Color? Data on African American, Hispanic, and Native American Faculty in STEM." *Journal of College Science Teaching*, 2010, *39*(4), 8–9.

Trenor, J. M., and others. "The Relations of Ethnicity to Female Engineering Students' Educational Experiences and College and Career Plans in an Ethnically Diverse Learning Environment." *Journal of Engineering Education*, 2008, *97*(4), 449–465.

DAWN R. JOHNSON is an assistant professor of the Higher Education Program at Syracuse University. Her research focuses on students of color in science and engineering, with a special interest in the experiences of women of color and their sense of belonging. She formerly directed recruitment and retention programs for underrepresented students at a science and engineering university.

NEW DIRECTIONS FOR INSTITUTIONAL RESEARCH • DOI: 10.1002/ir

The authors describe the scale development process for three new scales designed to measure attitudes and perceptions about scientists and offer implications for how these tools can be used on college campuses and for future research.

New Tools for Examining Undergraduate Students' STEM Stereotypes: Implications for Women and Other Underrepresented Groups

Sylvia C. Nassar-McMillan, Mary Wyer, Maria Oliver-Hoyo, Jennifer Schneider

Although both domestic U.S. and international statistics on population demographics within science, technology, engineering, and mathematics (STEM) fields indicate overall gains and more even representation among various groups, caution must be taken to interpret these gains as suggesting blanket improvement in underrepresentation issues. When data are dissected more closely, it becomes difficult to discern actual patterns, as the weight of evidence in recent decades is not necessarily consistent or easily interpretable. For example, on the one hand women have earned more STEM degrees across all levels in the last decade, including almost 40 percent of science and engineering degrees between 1997 and 2006 (see Figure 8.1). Correspondingly, non-white U.S. populations have increased their representation across STEM fields, including earning 49 percent of doctoral degrees in science and engineering between 1997 and 2006 (see Figure 8.2, noting that it includes noncitizens, thus distorting the actual number of U.S. citizens earning science and engineering doctoral degrees). On the other hand, these gains in representation among women and other underrepresented groups appear to be prevalent in specific STEM fields that may yield lower salaries or prestige within the world of work and other social structures, such as psychology and the social sciences, compared to fields such as physics, computer science, and engineering

NEW DIRECTIONS FOR INSTITUTIONAL RESEARCH, no. 152, Winter 2011 © Wiley Periodicals, Inc.
Published online in Wiley Online Library (wileyonlinelibrary.com) • DOI: 10.1002/ir.411

Figure 8.1. Population Representation in Science and Engineering Degrees

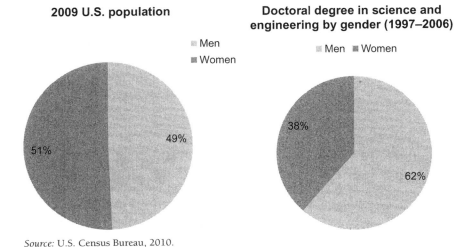

Source: U.S. Census Bureau, 2010.

Figure 8.2. Population Representation in Science and Engineering Degrees

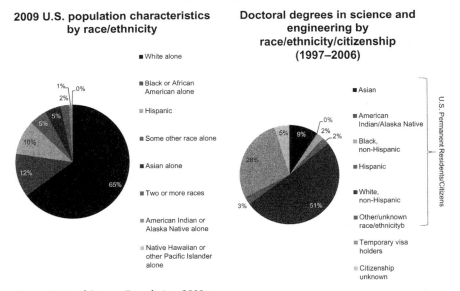

Source: National Science Foundation, 2009c.
Note: In doctoral degree chart, "Other/unknown race/ethnicity" includes Native Hawaiians/Other Pacific Islanders and multiple-race, non-Hispanic ethnicity, from 2001 forward.

(National Science Foundation, 2008, 2009a, 2009b). Thus there continues to be segregation within STEM participation in both degree level and specific STEM field.

These ongoing discrepancies are troubling on a number of levels. From an economic perspective, the United States is not able to sustain the supply needed either domestically or globally with its current pipeline of new trainees for STEM fields. From a social equity perspective, certain STEM fields tend to yield higher average salaries and lower unemployment than others, and these are the ones that, despite overall gains across nonwhite nonmale demographic groups in STEM fields, will continue to be economically and socially marginalized within the context of the current pipeline flow. Finally, and perhaps most importantly, federal mandates to create more equal access in educational and career opportunities are grounded in ethical principles of fairness and equality for all U.S. citizens alike. Underresourced school districts, "hostile" STEM climates in universities, and organizational work factors all play into the discourse of equity spanning educational, entry-level, and advanced career opportunities in STEM. These concerns, t ' en together, provide a strong rationale for continued policy initiatives to address the slowly changing landscape of STEM fields in the United States, as well as globally.

It is in this context of empirical inquiry that many scholars have attempted to identify the salient variables influencing individuals' career choice, particularly into STEM fields. One such variable is stereotyping and its relationship to occupational choice. Although these relationships have long been known to exist (Betz and Fitzgerald, 1987), the current work overviewed in this chapter focuses on this issue in a contemporary and culturally relevant context.

Career Stereotyping

Knowledge about both self and careers may have a concrete and factual basis, but it may also be rooted in individuals' stereotypes. These stereotypes may then serve to promote or inhibit creation of a conceptual match between the individual and a particular career.

People hold in their mind images of careers and the individuals they associate with those careers. Such images, or stereotypes, may manifest in attitudes toward such careers, serving to support or deter one from pursuing a specific career path. Stereotypes have a powerful impact for a variety of reasons (Crawford, 2006). From a personal perspective, they may become incorporated into one's self-concept and as such facilitate self-fulfilling prophecies. Moreover, they may reinforce differences in perceived status and power among self and others, provoking social dynamics such as discrimination or sexist behavior.

Prior research indicates differences among perceived images of scientists by ethnic and gender backgrounds of students from early elementary

NEW DIRECTIONS FOR INSTITUTIONAL RESEARCH • DOI: 10.1002/ir

age on up (Sumrall, 1995). Cross-cultural research also suggests differences among ethnic and gender groups in terms of their expressed interests in both career-related tasks as well as specific careers in general (Fouad, 2002; Fouad and Mohler, 2004). Historically, images of scientists have included white males wearing lab coats and glasses (Beardslee and O'Dowd, 1961; Chambers, 1983; Mead and Metraux, 1957; Newton and Newton, 1998), with characteristics spanning dedication, intelligence, and altruism to being uninteresting, unsociable, and dull (Chambers, 1983; Dikmenli, 2010; Lips, 1992; Mead and Metraux, 1957; Rahm and Charbonneau, 1997). Contemporary research findings suggest that science continues to be perceived as a masculine profession, with females subscribing more closely to that notion (Hughes, 2002; Lips, 1992; Thomas, Henley, and Snell, 2006), and conversely that males perceive scientists more positively than do females (Catsambis, 1995; Sax, 2001; Weinburgh, 1995).

In light of such research findings, it is not surprising that women and people of color are disproportionately underrepresented at the college and professional levels, particularly in fields such as the sciences and engineering.

It is not known whether women and people of color are actively discouraged from pursuing such professions or if their self-images and stereotypes exacerbate internalized beliefs about their own abilities with regard to science-related professions. Thus it is critical to examine the attitudes and stereotypes of women and people of color toward those fields as well as toward the professionals representing them, in order to identify process and practices that can lead to improving educational outcomes. Given the global rise in the number of women in the fields of interest here, alongside the context of uneven and perplexing gains (Glover, 2000), it is an opportune time to explore whether women and men have differing images of scientists, particularly since there is corresponding growth in popular media images of women as scientists and engineers (Barbercheck, 2001; Mendick, Moreau, and Hollingworth, 2008; Steinke, Long, Johnson, and Ghosh, 2008). If they do have relatively negative images compared to those of their male counterparts, the theory that women's stereotypes of scientists discourage their interest in STEM education will have continuing salience. Understanding those images is a necessary analytic step in advancing knowledge about the dynamics of gender in such education and in refining and tailoring the development of interventions.

The remainder of this chapter details the major findings from an innovative National Science Foundation project initially conceptualized to develop assessment tools to examine college students' images and attitudes toward science and scientists. The five-year project yielded three psychometrically sound instruments designed to examine levels of and relationships between individuals' images of scientists and self, career intentions in science, and perceptions of equality in science. These instruments are also designed to measure impacts of curricular or other interventions in science education. Moreover, they are currently being adapted, through a NASA-funded project, for other age groups and populations.

Description of Scale Development

With the goal of establishing content validity for our instrument throughout the process of its development, our steps involved (1) creating a pool of items based on a thorough literature review, existing scales, and our own experiences; (2) using seasoned experts by means of rigorous review and revision on the part of our experienced research team; and (3) consulting with participants. Although the ultimate goal of the project— creating a scale for measuring students' perceptions of scientists—represents an intrinsically quantitative design, we opted to employ several smaller-stage qualitative studies to develop content for our questionnaire and subsequent interventions (Ulin, Robinson, and Tolley, 2004). Concurrently with compiling relevant instrument items from the research literature, the early stages of this instrument development process involved use of focus groups to facilitate college students' discussion around salient influences on their views of science and scientists as well as their own career development in terms of decisions about pursuing science-related careers. This aspect of the process ensured that our resultant draft of the scale was directly informed by knowledge gained from the focus groups— in effect, reflecting the variety of perceptions students may have about scientists and how they interpret their realities (Racher and Robinson, 2002). A particular advantage to this format is that participants' comments and debate within the discussions yielded insight as to the contextual or social environment in which they were approaching their career decisions (Nassar-McMillan, Wyer, Oliver-Hoyo, and Ryder-Burge, 2010; Ulin, Robinson, and Tolley, 2004).

After two rounds of focus groups were completed, a pilot administration was conducted at a large, southeastern research extensive institution with more than 1,100 undergraduate participants. This stage refined the instrument to the point at which it was ready to be administered nationally for final psychometric validation as well as to do initial hypothesis testing among the variables. The final administration was conducted across the United States, with a total of 1,639 undergraduate university students from seven universities representing various geographic regions (42 percent southeastern United States, 20 percent midwestern, 20 percent southern, and 18 percent northeastern). Women represented 59 percent of the sample; men represented 41 percent. Student participants, recruited from large undergraduate introductory science courses taught by instructors identified through PI's national science and engineering networks, were from a broad range of majors, with the largest being agricultural and life science (48 percent) and the smallest being social science (6 percent), other (4 percent), or undecided (3 percent). A description of the three scales is outlined below.

Stereotypes of Scientists (SOS) Scale. The primary set of items, focusing on stereotyping in science, was drawn from instruments such as the Images of Scientists and Science Scale or ISSS (Krajkovich and Smith, 1982; She, 1992) and the Women in Science Scale or WiSS (Erb and Smith, 1984; Owen and others, 2007). Additionally, the focus group process used the Draw-A-Scientist-Test or DAST (Chambers, 1983; Rahm and Charbonneau, 1997) as a framework for stimulating discussion around stereotypes in science. These findings were confirmed in a national study that resulted in a slightly reduced item set that yielded a scale with eighteen items and two factors (with each yielding nine items).

An exploratory factor analysis with promax rotation was conducted using the original items from the Stereotypes of Scientists Scale (from both pilot study and final administration stages). A slightly reduced item set then underwent a confirmatory factor analytic procedure, with the final scale yielding eighteen items and two factors (each with nine items). The Professional Competencies factor (Cronbach's alpha = .81) items are worded with the sentence stem "I think that scientists are . . ." followed by items such as "especially intelligent" and "able to learn to use new equipment quickly." The Interpersonal Competencies factor (alpha = .77) consists of items such as "I think that scientists are . . ." followed by descriptors such as "family-oriented" and "cooperative." Likert scale response options were denoted by 1 = strongly disagree to 6 = strongly agree.

Career Intentions in Science (CIS) Scale. Initial items on the CIS scale were informed by previous research, as well as occupational choice and motivation (Ellis and Herrman, 1983), career aspirations and ambitions in science (Ellis and Herrman, 1983; Schoon, 2001), and self-efficacy in science (Lent Brown, and Hackett, 1994; McLean and Kalin, 1994), with specific scales examined including the Self Directed Search or SDS (Holland, 1994), Strong Interest Inventory or SII (Donnay and Borgen, 1996; Lattimore and Borgen, 1999; Fouad, 2002; Fouad and Mohler, 2004), and the Career Decision Profile (Jones and Lohmann, 1998).

Again, exploratory and confirmatory factor analyses were conducted, reducing twenty-six initial items to a dozen items making up a single factor (alpha = .98) with factor loadings ranging from .81 to .92. These items, also rated via Likert scale (1 = very unlikely, 6 = very likely), posed the question stem "In your future career, how likely is it that you will . . ." followed by items such as "get an advanced degree in science" and "have a very successful career in science."

Social Equality Perceptions in Science (SEPS) Scale. The third scale was initially informed by previous scales measuring various aspects and perceptions of social equality. The Feminist Identity Development Scale or FIDS (Bargad and Hyde, 1991), Feminist Identity Scale or FIS

(Fischer and others, 2000), Feminist Identity Composite or FIC (Fischer and others, 2000), Womanist Identity Attitudes Scale or WIDS (Moradi, 2004), Attitudes Towards Women Scale or AWS (Spence and Hahn, 1997), Modern Sexism or MS (Swim, Aiken, Hall and Hunter, 1995), Ambivalent Sexism Inventory or ASI (Glick and Fiske, 1997), Feminist Perspectives Scale or FPS (Henley and others, 1998), and the Bem Sex Role Inventory or BSRI (Harris, 1994) were examined for prospective items for inclusion in the SEPS Scale. Our review of these scales resulted in adding a set of items that distinguished perceptions of current environment from perceptions about the ideal environment.

The factor analyses yielded two strong factors of fourteen total items. The two subscales, ideal science opportunities (alpha = .93) and actual science opportunities (alpha = .92), were made up of companion items such as (for the ideal subscale) "Please rate the extent to which you agree with the following statements: people of all ethnic groups should receive equal educational opportunities in science; women and men should receive equal employment opportunities in science" with (for the actual subscale) "Please rate the extent to which you agree with the following statements: People of all ethnic groups do receive equal employment opportunities in science; women and men do receive equal employment opportunities in science." The overall SEPS scale score was calculated by computing the discrepancy between average companion ideal and actual item scores, with high SEPS scale scores denoting large discrepancies between perceived ideal and actual opportunities across STEM education and careers.

Preliminary Hypotheses Testing

A few additional questions were added to the demographic and background questionnaire portion of the overall instrument. Even though the data yielded by these item responses have not yet been empirically examined, the team thought it important to add items to query participants about their own personal and cultural life experiences, on the basis of issues emergent from the initial focus groups. Instruments such as the Multigroup Ethnic Identity Measure or MEIM (Worrell, 2000) and Vancouver Index of Acculturation or VIA (Ryder, Alden and Paulhus, 2000) informed our inclusion of items such as "I am comfortable working with members of different ethnic groups" and "I have family members who are scientists."

Although the initial purpose of the project was to focus on psychometric support for the instruments developed, a few hypotheses were also tested preliminarily that warrant further investigation. With regard to findings from the SOS scores, we found, in both pilot and national administrations, that participant STEM majors had higher mean scores on stereotypes regarding interpersonal competences than did their non-STEM counterparts (in effect, STEM majors attributed higher interpersonal competencies than did non-STEM majors). Moreover, higher mean scores on the two factors (interpersonal; professional competencies) were related to scores on the Career Intentions in Science Scale, as determined through a multiple regression, F (17, 1621) = 85.40, $p < .001$. Although these results are interesting and contribute new empirical insights about issues of stereotyping in science, they are most useful because they offer a measure of predictive validity for the scales.

Another set of findings reveals differences of perspectives held by gender and race or ethnicity. On the Social Equality Perceptions Scale scores, people from underrepresented groups were significantly less likely than whites to agree that things are (i.e., actually) equal ($F = 56.78$, $p < .01$); women are more likely than men to agree that things should (i.e., ideally) be equal ($F = 76.21, p < .01$).

Implications for Research and Practice

Clearly, exposure to stereotypes occurs much earlier than during undergraduate years. Although examining college students' stereotypes about STEM fields is critical in order to inform university policy and practice, assessments need to be conducted at earlier points in time as well. The degree to which these stereotypes may matter, or not, changes as people develop their career interests and advance their education. This is a critical point and leads to an important level of sophistication and complexity in that these new scales offer researchers empirical tools to evaluate changes over time—not just in stereotypes but in relationship of stereotypes to other elements of educational persistence and career choice. Moreover, inquiry into individuals' self-perceptions, as well as their perceptions of others, including scientists, will help determine how matches or discrepancies between perceptions of self and others influence career choice. Further, demographic and other background variables, such as familial and community social structures, also need to be examined in relation to career choice. Concurrent cross-cultural inquiry would help to inform domestic issues within a broader and more appropriate global context, thus prescribing cross-cultural adaptations and norming of similar instruments.

Such inquiry can be particularly useful if testing the impacts of curriculum interventions designed to address issues of stereotyping, particularly in STEM fields. At the same time, similar inquiry across a broader range of STEM fields is necessary and timely. All of these initiatives will inform the testing of modified models of educational and career development by incorporating stereotyping into contemporary, culturally relevant models—for example, Self-Efficacy Career Theory (Betz and Hackett, 1997) and Social Cognitive Career Theory (Lent, Brown, and Hackett, 1994), two theories discussed earlier in this volume and noted for their relevance to a broad range of demographic groups.

In terms of practice, some of our premises and findings are helpful to keep in mind while working on or with college campuses. Whenever possible, it is important to counter negative views (e.g., stereotypes) of science and scientists. Although our results trended toward "positive" stereotypes, particularly among STEM majors, the variance among demographic groups is important to remember. Thus, it is essential to encourage, rather than discourage, students of all ages and demographic groups alike in STEM disciplines and subject matter. In this way, we can positively influence students' self-efficacy.

Psychosocially, we need to understand that relationships do exist between stereotypes and students' career choices. Moreover, we must acknowledge disparities in equity perceptions among various demographic groups, often relative to their representation or underrepresentation in STEM preparation, majors, or fields. We need to examine our curriculum interventions both quantitatively and qualitatively. For example, in addition to setting up adequate pre- and post-examination designs within our classroom environments, we should create and harness "teachable" moments whenever possible. We have to be willing to discuss attitudes, their precursors, and their evolutions among the student groups we serve. Above all, we can examine our own career development processes in the context of our gender and cultural or ethnic life experiences. After all, it is difficult—if not impossible—to teach what we do not yet know ourselves.

References

Barbercheck, M. "Mixed Messages: Men and Women in Advertisements in Science." In M. Wyer, M. Barbercheck, D. Geisman, H. O. Ozturk, and M. Wayne (eds.), *Women, Science, and Technology: A Reader in Feminist Science Studies.* New York: Routledge, 2001.

Bargad, A., and Hyde, J. S. "Women's Studies: A Study of Feminist Identity Development in Women." *Psychology of Women Quarterly,* 1991, *15*(2), 181–201.

Beardslee, D. C., and O'Dowd, D. "The College-Student Image of the Scientist." *Science,* 1961, *133*(3457), 997–1001.

Betz, N. E., and Fitzgerald, L. F. *The Career Psychology of Women.* San Diego, Calif.: Academic Press, 1987.

Betz, N., and Hackett, G. "Applications of Self-Efficacy Theory to the Career Development of Women." *Journal of Career Assessment,* 1997, *5*, 383–402.

Catsambis, S. "Gender, Race, Ethnicity, and Science Education in the Middle Grades." *Journal of Research in Science Teaching,* 1995, *32*, 243–257.

Chambers, D. W. "Stereotypic Images of the Scientists: The Draw-A Scientist Test." *Science Education,* 1983, *67*, 255–265.

Crawford, M. *Transformations: Women, Gender and Psychology.* New York: McGraw-Hill: 2006.

Dikmenli, M. "Undergraduate Biology Students' Representations of Science and the Scientist." *College Student Journal,* 2010, *44*(2), 579–588.

Donnay, D.A.C., and Borgen, F. H. "Validity, Structure, and Content of the 1994 Strong Interest Inventory." *Journal of Counseling Psychology,* 1996, *43*(3), 275–291.

Ellis, R. A., and Herrman, M. S. "Three Dimensions of Occupational Choice: A Research Note on Measuring the Career Intentions of College Women." *Social Forces*, 1983, *63*(3), 893–903.

Erb, T. O., and Smith, W. S. "Validation of the Attitude Toward Women in Science Scale for Early Adolescents." *Journal of Research in Science Teaching*, 1984, *21*(4), 391–397.

Fischer, A. R., and others. "Assessing Women's Feminist Identity Development." *Psychology of Women Quarterly*, 2000, *24*(1), 15–29.

Fouad, N. A. "Cross-Cultural Differences in Vocational Interests: Between Group Differences on the Strong Interest Inventory." *Journal of Counseling Psychology*, 2002, *49*, 282–289.

Fouad, N. A., and Mohler, C. J. "Cultural Validity of Holland's Theory and the Strong Interest Inventory for Five Racial/Ethnic Groups." *Journal of Career Assessment*, 2004, *12*, 423–439.

Glick, P., and Fiske, S. T. "The Ambivalent Sexism Inventory: Differentiating Hostile and Benevolent Sexism." *Journal of Personality and Social Psychology*, 1997, *70*(3), 491–512.

Glover, J. *Women and Scientific Employment*. Basingstoke, England: Macmillan, 2000.

Harris, A. C. "Ethnicity as a Determinant of Sex Role Identity: A Replication Study of Item Selection for the Bem Sex Role Inventory." *Sex Roles*, 1994, *31*(3/4), 241–253.

Henley, N. M., and others. "Developing a Scale to Measure the Diversity of Feminist Attitudes." *Psychology of Women Quarterly*, 1998, *22*(3), 317–348.

Holland, J. L. *SDS: Self Directed Search, Form R, 4th Edition, Assessment Booklet*. Psychological Assessment Resources, 1994.

Hughes, W. J. "Gender Attributions of Science and Academic Attributes: An Examination of Undergraduate Science, Mathematics, and Technology Majors." *Journal of Women and Minorities in Science and Engineering*, 2002, *8*, 53–65.

Jones, L. K., and Lohmann, R. C. "The Career Decision Profile: Using a Measure of Career Decision Status in Counseling." *Journal of Career Assessment*, 1998, *6*(2), 209–230.

Krajkovich, J. G., and Smith, J. K. "The Development of the Image of Science and Scientists Scale." *Journal of Research in Science Teaching*, 1982, *19*(1), 39–44.

Lattimore, R. R., and Borgen, F. H. "Validity of the 1994 Strong Interest Inventory with Racial and Ethnic Groups in the United States." *Journal of Counseling Psychology*, 1999, *46*(2), 185–195.

Lent, R. W., Brown, S. D., and Hackett, G. "Toward a Unifying Social Cognitive Theory of Career and Academic Interest, Choice, and Performance." *Journal of Vocational Behavior*, 1994, *45*, 79–122.

Lips, H. M. "Gender- and Science-Related Attitudes as Predictors of College Students." *Journal of Vocational Behavior*, 1992, *40*(1), 62–81.

Moradi, B. "An Evaluation of the Psychometric Properties of the Womanist Identity Attitudes Scale." *Sex Roles*, 2004, *50*(3/4), 253–266.

McLean, H. M., and Kalin, R. "Congruence Between Self-Image and Occupational Stereotypes in Students Entering Gender-Dominated Occupations. *Canadian Journal of Behavioural Science*, 1994, *26*(1), 142–162.

Mead, M., and Metraux, R. "The Image of the Scientist Among High-School Students." *Science*, 1957, *126*(3270), 384–390.

Mendick, H., Moreau, M., and Hollingworth, S. *Mathematical Images and Gender Identities: A Report on the Gendering of Representations of Mathematics and Mathematicians in Popular Culture and Their Influences on Learners*. Bradford: UK Resource Centre for Women in Science Engineering and Technology, 2008.

Nassar-McMillan, S. C., Wyer, M., Oliver-Hoyo, M., and Ryder-Burge, A. "Using Focus Groups in Instrument Development: Expected and Unexpected Lessons Learned." *The Qualitative Report*, 2010, *15*(6), 1621–1634.

National Science Foundation, Division of Science Resources Statistics. "Science and Engineering Degrees: 1966–2006." Detailed Statistical Tables NSF 08-321. Arlington, Va.: National Science Foundation, 2008. Retrieved Feb. 1, 2010, from http://www.nsf.gov/statistics/nsf08321/.

National Science Foundation, Division of Science Resources Statistics (NSF). "Doctorate Recipients from U.S. Universities: Summary Report 2007–2008." Special Report NSF 10-309. Arlington, Va.: National Science Foundation, 2009a. Retrieved Feb. 1, 2010, from http://www.nsf.gov/statistics/nsf10309/.

National Science Foundation, Division of Science Resources Statistics (NSF). "Women, Minorities, and Persons with Disabilities in Science and Engineering: 2009." NSF 09-305. Arlington, Va.: National Science Foundation, 2009b. Retrieved Feb. 1, 2010, from http://www.nsf.gov/statistics/nsf10309/.

National Science Foundation, Division of Science Resources Statistics (NSF). "2010 Science and Engineering Degrees, by Race/Ethnicity of Recipients: 1997–2006." Detailed Statistical Tables NSF 10-300. Arlington, Va.: National Science Foundation, 2009c. Retrieved November 29, 2011, from http://www.nsf.gov/statistics/nsf10300/.

Newton, L. D., and Newton, D. P. "Primary Children's Conceptions of Science and the Scientist: Is the Impact of a National Curriculum Breaking Down the Stereotype?" International Journal of Science Education, 1998, 20(9), 1137–1149.

Owen, S. V., and others. "Psychometric reevaluation of the Women in Science Scale (WiSS)." Journal of Research in Science Teaching, 2007, 44(10), 1461–1478.

Racher, F. E., and Robinson, S. "Are Phenomenology and Postpositivism Strange Bedfellows?" Western Journal of Nursing Research, 2002, 25, 464–481.

Rahm, J., and Charbonneau, P. "Probing Stereotypes Through Student's Drawings of Scientists." American Journal of Physics, 1997, 65(8), 774–778.

Ryder, A. G., Alden, L. E., and Paulhus, D. L. "Is Acculturation Unidimensional or Bidimensional? A Head-to-Head Comparison in the Prediction of Personality, Self-Identity, and Adjustment." Journal of Personality and Social Psychology, 2000, 79(1), 49–65.

Sax, L. J. "Undergraduate Science Majors: Gender Differences in Who Goes to Graduate School." Review of Higher Education, 2001, 24(2), 153–172.

She, H.-C. The Impact of a Biochemistry Workshop on Gifted Children's Image of Science and Scientists, Women in Science and Class Participation. Unpublished doctoral dissertation, University of Missouri, Columbia, 1992.

Schoon, I. "Teenage Job Aspirations and Career Attainment in Adulthood: A 17 Year Follow-Up Study of Teenagers Who Aspired to Become Scientists, Health Professionals, or Engineers." International Journal of Behavioral Development, 2001, 25, 124–132.

Spence, J. T., and Hahn, E. D. "The Attitudes Toward Women Scale and Attitude Change in College Students." Psychology of Women Quarterly, 1997, 21, 17–34.

Sumrall, W. "Reasons for the Perceived Images of Scientists by Race and Gender of Students in Grades 1–7." School Science and Mathematics, 1995, 95, 83–90.

Steinke, J., Long, M., Johnson, M. J., and Ghosh, S. "Gender Stereotypes of Scientist Characters in Television Programs Popular Among Middle School–Aged Children." Paper presented to the Science Communication Interest Group (SCIGroup) at the Annual Meeting of the Association for Education in Journalism and Mass Communication (AEJMC), Chicago, August 2008.

Swim, J. K., Aiken, K. J., Hall, W. S., and Hunter, B. A. "Sexism and Racism: Old-Fashioned and Modern Prejudices." Journal of Personality and Social Psychology, 1995, 68, 199–214.

Thomas, M. D., Henley, T. B., and Snell, C. M. "Draw a Scientist Test: A Different Population and a Somewhat Different Story." College Student Journal, 2006, 40(1), 140–148.

Ulin, P. R., Robinson, E. T. and Tolley, E. E. "Qualitative Data Analysis." In P. R. Ulin, E. T. Robinson, and E. E. Tolley (eds.), Qualitative Methods in Public Health: A Field Guide for Applied Research. San Francisco: Jossey-Bass, 2004.

U.S. Census Bureau. "Table 3. Annual Estimates of the Resident Population by Sex, Race, and Hispanic Origin for the United States: April 1, 2000 to July 1, 2009," NC-EST2009-03. Washington, D.C.: U.S. Census Bureau, 2010. Retrieved November 29, 2011, from http://www.census.gov/popest/national/asrh/2009-nat-detail.html.

Weinburgh, M. "Gender Differences in Student Attitudes Toward Science: A Meta-Analysis of the Literature from 1970 to 1991." *Journal of Research in Science Teaching*, 1995, *32*(4), 387–398.

Worrell, F. C. "A Validity Study of Scores on the Multigroup Ethnic Identity Measure Based on a Sample of Academically Talented Adolescents." *Educational and Psychological Measurement*, 2000, *60*, 3, 439–447.

SYLVIA NASSAR-MCMILLAN is Professor and Program Coordinator of Counselor Education at North Carolina State University.

MARY WYER is Associate Professor of Psychology and Women's and Gender Studies at North Carolina State University.

MARIA OLIVER-HOYO is Associate Professor of Chemistry at North Carolina State University.

JENNIFER SCHNEIDER is Assistant Director of Diversity Research and Assessment at the University of North Texas, Denton.

INDEX

external accountability purposes. This volume of *New Directions for Institutional Research* makes a compelling case that institutions can and should be assessing consequential, complex general education student learning outcomes. It also gives faculty members and assessment leaders the tools and resources to take ownership of this important work. Part One of this volume provides an argument for why we should be assessing general education and describes a framework, based on a rigorous psychological research approach, for engaging in assessment. The six chapters in Part Two show how this work can be (and is being) done for six important learning outcomes: critical thinking, quantitative reasoning, teamwork, intercultural competence, civic knowledge and engagement, and integrative learning. The volume closes with recommendations on needed innovations in general education assessment and presents a research agenda for future work.
ISBN: 978-1-1180-9133-3

IR 148   **Students of Color in STEM**
Shaun R. Harper, Christopher B. Newman
Why are some racial minorities so underrepresented as degree candidates in science, technology, engineering, and mathematics (STEM)? Why are they so underprepared for college-level math and science courses? Why are their grades and other achievement indicators disproportionately lower than their white counterparts? Why do so many of them change their majors to non-STEM fields? And why do so few pursue graduate degrees in STEM? These five questions are continuously recycled in the study of students of color in STEM. Offered in this volume of *New Directions for Institutional Research* are new research ideas and frameworks that have emerged from recent studies of minorities in STEM fields across a wide array of institution types: large research universities, community colleges, minority-serving institutions, and others. The chapter authors counterbalance examinations of student underperformance and racial disparities in STEM with insights into the study of factors that enable minority student success.
ISBN: 978-1-1180-1402-8

IR 147   **System Offices for Community College Institutional Research**
Willard C. Hom
This volume of *New Directions for Institutional Research* examines a professional niche that tends to operate with a low profile while playing a major role in state policies—the system office for community college institutional research. As states, regions, and the federal government seek ways to evaluate and improve the performance of community colleges, this office has grown in importance. The chapter authors, all institutional researchers in this area, draw a timely state-of-the-niche portrait by showing how this office varies across states, how it varies from other institutional research offices within states, and the implications its history and prospects have for the future. This volume will be particularly useful for those who deal with higher education policy at the state, regional, or federal level; on-campus institutional researchers; and individuals who currently work in or with these system offices.
ISBN: 978-04709-39543

IR 146   **Institutional Research and Homeland Security**
Nicolas A. Valcik
Although homeland security has captured the public's attention in recent years, higher education institutions have had to contend with emergency situations and security issues long before 9/11 occurred. Well known incidents such as the Unabomber attacks and decades of sporadic school shootings brought violence to college campuses long before the Department

of Homeland Security was established. Despite these past security issues and the passage of the PATRIOT Act, very little research has been performed on homeland security issues and higher education institutions. This volume of *New Directions for Institutional Research* examines how new federal regulations impact institutional research and higher education institutions. This volume also addresses key issues such as right-to-privacy regulations, criminal background checks, the Student and Exchange Visitor Information System (SEVIS), information technology security, the use of geographic information systems as a research tool, hazardous materials (HAZMAT) management, and the impact of natural disasters and manmade threats on applications and enrollment.
ISBN: 978-04709-03148

IR 145 **Diversity and Educational Benefits**
Serge Herzog
Campus climate studies and research on the impact of diversity in higher education abound. On closer examination, however, the corpus of findings on the role of diversity and how diversity is captured with campus climate surveys reveals both conceptual and methodological limitations. This volume of *New Directions for Institutional Research* addresses these limitations with the inclusion of studies by institutional research (IR) practitioners who make use of data that furnish new insights into the relationships among student diversity, student perception of campus climate, and student sociodemographic background—and how those relationships affect academic outcomes. Each chapter emphasizes how IR practitioners benefit from the conceptual and analytical approach laid out, and each chapter provides a framework to gauge the contribution of diversity to educational benefits. The findings revealed in this volume cast doubt on the benefits of student diversity purported in previous research. At a minimum, the influence of student diversity is neither linear nor unidirectional, but operates within a complex web of inte. .lated factors that shape the student experience.
ISBN: 978-04707-67276

IR 144 **Data-Driven Decision Making in Intercollegiate Athletics**
Jennifer Lee Hoffman, James Soto Antony, Daisy D. Alfaro
Data related to intercollegiate athletics are often a small part of campus financial and academic data reporting, but they generate significant interest at any institution that sponsors varsity sports. The demands for documentation, accountability, and data-driven decision making related to college athletics have grown increasingly sophisticated. These demands come from the press, campus decision makers, researchers, state and federal agencies, the National Collegiate Athletic Association, and the public. Despite the growth of data sources and the ease of access that information technology affords, gaps still exist between what we think we know about college athletics and supporting data. The challenge for institutional researchers is to continue developing consistent data sources that inform the policy and governance of college athletics. This volume of *New Directions for Institutional Research* introduces the reader to the primary and secondary sources of data on college athletics and their utility for decision making. The authors describe the existing landscape of data about student athletes and intercollegiate athletics and the measures that are still needed.
ISBN: 978-04706-08289

NEW DIRECTIONS FOR INSTITUTIONAL RESEARCH
ORDER FORM SUBSCRIPTION AND SINGLE ISSUES

DISCOUNTED BACK ISSUES:

Use this form to receive 20% off all back issues of *New Directions for Institutional Research*.
All single issues priced at **$23.20** (normally $29.00)

TITLE	ISSUE NO.	ISBN

Call 888-378-2537 or see mailing instructions below. When calling, mention the promotional code JBNND to receive your discount. For a complete list of issues, please visit www.josseybass.com/go/ndir

SUBSCRIPTIONS: (1 YEAR, 4 ISSUES)

☐ New Order   ☐ Renewal

U.S.	☐ Individual: $109	☐ Institutional: $297
CANADA/MEXICO	☐ Individual: $109	☐ Institutional: $337
ALL OTHERS	☐ Individual: $133	☐ Institutional: $371

Call 888-378-2537 or see mailing and pricing instructions below.
Online subscriptions are available at www.onlinelibrary.wiley.com

ORDER TOTALS:

Issue / Subscription Amount: $ _____

Shipping Amount: $ _____
(for single issues only – subscription prices include shipping)

Total Amount: $ _____

SHIPPING CHARGES:	
First Item	$6.00
Each Add'l Item	$2.00

(No sales tax for U.S. subscriptions. Canadian residents, add GST for subscription orders. Individual rate subscriptions must be paid by personal check or credit card. Individual rate subscriptions may not be resold as library copies.)

BILLING & SHIPPING INFORMATION:

☐ **PAYMENT ENCLOSED:** *(U.S. check or money order only. All payments must be in U.S. dollars.)*

☐ **CREDIT CARD:**   ☐ VISA   ☐ MC   ☐ AMEX

Card number _____ Exp. Date _____

Card Holder Name _____ Card Issue # _____

Signature _____ Day Phone _____

☐ **BILL ME:** *(U.S. institutional orders only. Purchase order required.)*

Purchase order # _____
Federal Tax ID 13559302 • GST 89102-8052

Name _____

Address _____

Phone _____ E-mail _____

Copy or detach page and send to: **John Wiley & Sons, One Montgomery Street, Suite 1200, San Francisco, CA 94104-4594**

Order Form can also be faxed to: **888-481-2665**

PROMO JBNND

Statement of Ownership

Statement of Ownership, Management, and Circulation (required by 39 U.S.C. 3685), filed on OCTOBER 1, 2011 for NEW DIRECTIONS FOR INSTITUTIONAL RESEARCH (Publication No. 0271-0579), published Quarterly for an annual subscription price of $109 at Wiley Subscription Services, Inc., at Jossey-Bass, One Montgomery St., Suite 1200, San Francisco, CA 94104-4594.

The names and complete mailing addresses of the Publisher, Editor, and Managing Editor are: Publisher, Wiley Subscription Services, Inc., A Wiley Company at San Francisco, One Montgomery St., Suite 1200, San Francisco, CA 94104-4594; Editor, Paul Umbach, North Carolina State University, Raleigh, NC 27695; Managing Editor, Robert Rosenberg, Wiley Subscription Services Inc., One Montgomery St., Suite 1200, San Francisco, CA 94104-4594. Contact Person: Joe Schuman; Telephone: 415-782-3232.

NEW DIRECTIONS FOR INSTITUTIONAL RESEARCH is a publication owned by Wiley Subscription Services, Inc.,111 River St., Hoboken, NJ 07030. The known bondholders, mortgagees, and other security holders owning or holding 1% or more of total amount of bonds, mortgages, or other securities are(see list).

	Average No. Copies Each Issue During Preceding 12 Months	No. Copies Of Single Issue Published Nearest To Filing Date (Summer 2011)		Average No. Copies Each Issue During Preceding 12 Months	No. Copies Of Single Issue Published Nearest To Filing Date (Summer 2011)
a. Total number of copies (net press run)	930	799	15d(2). In-county nonrequested copies stated on PS form 3541	0	0
b. Legitimate paid and/or requested distribution (by mail and outside mail)			15d(3). Nonrequested copies distributed through the USPS by other classes of mail	0	0
b(1). Individual paid/requested mail subscriptions stated on PS form 41 (include direct written request from recipient, telemarketing, and internet requests from recipient, paid subscriptions including nominal rate subscriptions, advertiser's proof copies, and exchange copies)	301	251	15d(4). Nonrequested copies distributed outside the mail	0	0
			15e. Total nonrequested distribution (sum of 15d(1), (2), (3), and (4))	56	55
			15f. Total distribution (sum of 15c and 15e)	357	306
b(2). Copies requested by employers for distribution to employees by name or position, stated on PS form 3541	0	0	15g. Copies not distributed	573	493
			15h. Total (sum of 15f and 15g)	930	799
b(3). Sales through dealers and carriers, street vendors, counter sales, and other paid or requested distribution outside USPS through USPS	0	0	15i. Percent paid and/or requested circulation (15c divided by 15f times 100)	84.3%%	82.0%%
b(4). Requested copies distributed by other mail classes	0	0	I certify that all information furnished on this form is true and complete. I understand that anyone who furnishes false or misleading information on this form or who omits material or information requested on this form may be subject to criminal sanctions (including fines and imprisonment) and/or civil sanctions (including civil penalties).		
c. Total paid and/or requested circulation (sum of 15b(1), (2), (3), and (4))	301	251	Statement of Ownership will be printed in the Winter 2011 issue of this publication.		
d. Nonrequested distribution (by mail and outside mail)					
d(1). Outside county nonrequested copies stated on PS form 3541	56	55	(signed) Susan E. Lewis, VP & Publisher-Periodicals		